T0276173

REVELATION OF JESUS CHRIST: DECODED

By the Standard Model of particle physics, energy, mass and celerity^2. Decoded over 12 years; between April 2007 | August 2019.

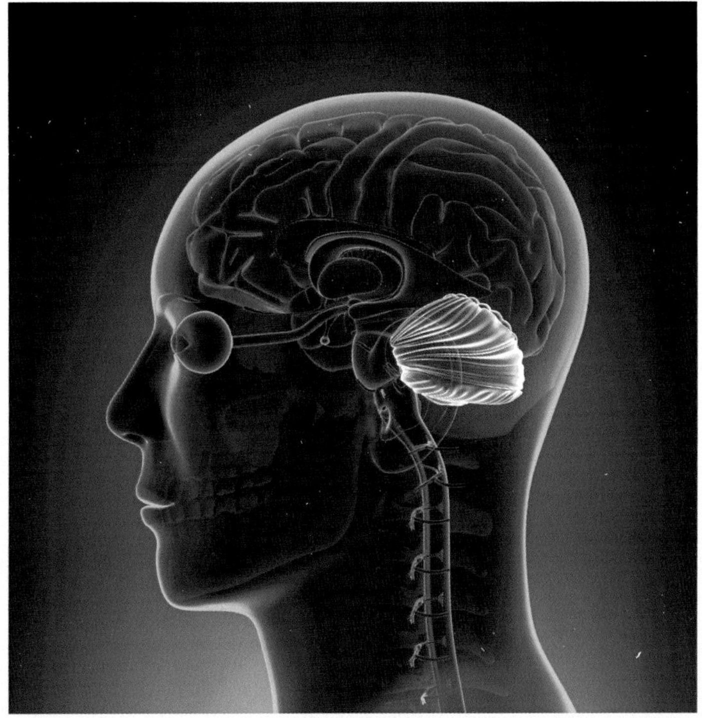

The human brain: Jesus Christ is the light of the world. The world the human anatomy; in the hindbrain. He is the sun, the nuclear reactor that lights up the anatomical systems within us that gives us life. He rules over the forces that move our limbs, the masses within us and the light that gives us intelligence. His throne is the central nervous system, the cerebellar cortex. At the midbrain are gluons, the fortifier.

Connie J. Allen

Constance, the woman who fought the dragon and lived to tell about it.

To order additional copies of this book, contact:
Xlibris
1-888-795-4274
www.Xlibris.com
Orders@Xlibris.com

ISBN: Softcover 978-1-7960-6163-5
EBook 978-1-7960-6162-8

Print information available on the last page

Rev. date: 10/03/2019

Children of God

Michelangelo, Last Judgment center detail (1541)

Of these three Jewels, Judah in the center as Christ and the woman, Israel is below him behind the muscle-ripped strong nuclear force, on the right is Jerusalem behind the electroweak or weak force. The weak force has shed his electroweak force mask and is offering electromagnetism's food, fecal matter and urine knitted as seminal fluids to Christ and the woman.

Only the three in the center panel of the Last Judgment matter. Skin color, wealth, social standing, beauty, education and religious affiliation have no bearing on who or what you are. Nothing matters except for these three. We, who are called "Jewels," are one of these three. The Gentiles, quarks and antiquarks are not in the center panel. Their names are not in the Lamb's book of life from before the world was made.

Contents

Revelation 1: Jesus Christ, a coded message in a homophone

Introduction

1 The Revelation of Jesus Christ,

> Jesus Christ, a homophone: He's us, crystallized from the hydrogen-based gluon soup. Jesus Christ is the beginning point of energy, mass and light, every person

which God gave unto him,

> God: Atom Lambda, of the Big Bang fame and fortune, the beginning of all things visible, the universe.

to shew unto his servants things which must shortly come to pass; and he sent and signified it by his angel,

> the Z or W^3 boson

unto his servant John,

> the Gentiles, a son of Abraham, conceived by those in the unified group, SU(3), the Z boson, photon and the W^3 boson;

2 Who bare record of the word of God,

> Jesus Christ is the word; Atom Lambda is God;

and of the testimony of Jesus Christ; and

> John was chosen to record

of all things that he saw.

3 Blessed is

> Jesus Christ,

he that readeth, and they that hear the words of this prophecy,

> in their minds,

and keep those things

> in their understanding,

which are written therein: for the time

> for the return of Jesus Christ to the minds of all people

is at hand.

Michelangelo, Creation of Adam (1541)

In the center panel, Adam sits on a rock, demonstrating his crystallization. Adam's hand is touching the hand of Jesus Christ whose hand provides life. Jesus Christ resides in the hindbrain, Adam's brain. Behind Jesus Christ are particles of mass, Judah, Israel and Jerusalem, and particles of light. Beneath the arm of Christ is a gluon, the archangel Michael. Below the right thigh with a stream of green, lethal waste, fecal matter and urine, is Jerusalem, the particle, force that is unified with electromagnetism. Every person is affiliated with Christ in this way. We live because he lives within us.

Seven churches in Asia

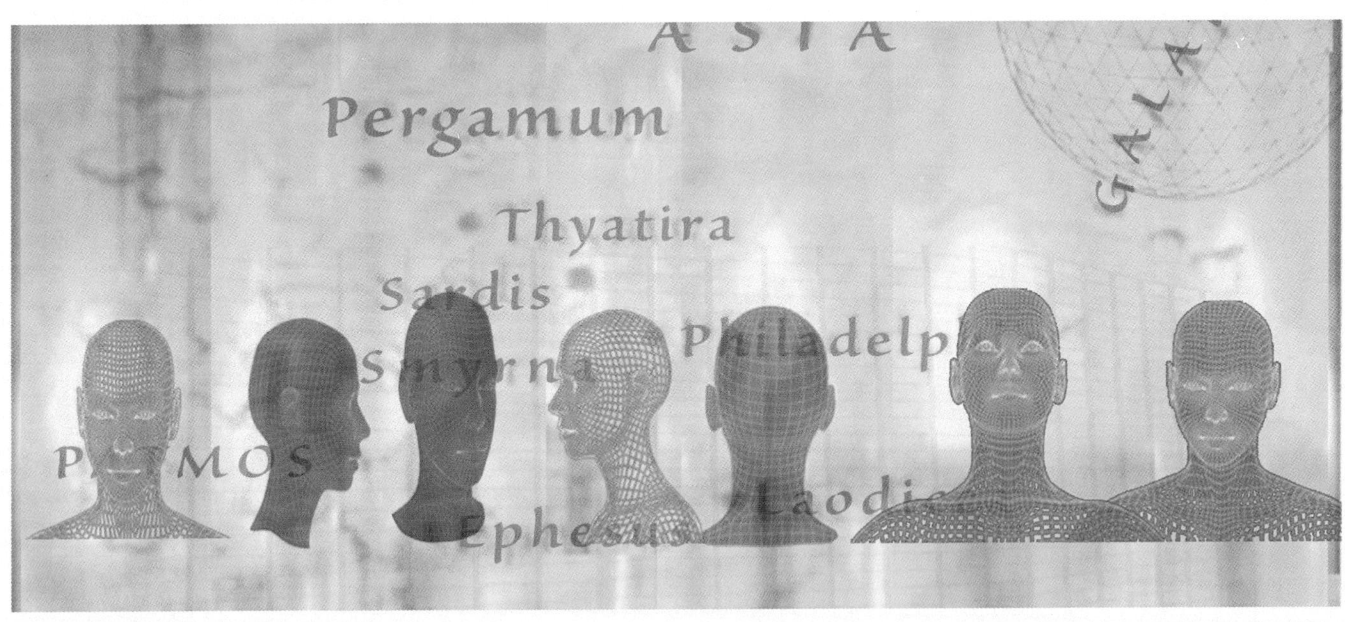

Seven Churches

Asia is located in the east at the midbrain. The heads are lamps, vehicles for thinking, lit by Jesus Christ. The human anatomy is lampstands, where the minds think.

4 John to the seven churches

A church is a complete body, a composition of energy, mass and light, that are mediated by a single leader,

which are in Asia:

Asia is in the east, the birth place of the mind; where wise men are born and where more wisdom is provided.

Grace be unto you,

leniency

and peace,

freedom from electricity, the devil and magnetism, otherwise known as Satan;

from him

Jesus Christ, a proton particle,

which is

alive,

and which was,

dead, murdered from the foundation of the world; crucified because of fear, electromagnetism

and which is to come,

returning to his rightful place as ruler over all energy, mass and light within every living body;

and from the seven Spirits

consisting of Higgs bosons (male and a separate female), gluons (male and a separate female), a Z boson (a female) and the W^3 boson (a male) tethered as one; a photon carrying electricity conjoined with a photon carrying magnetism,

which are

in the midbrain

before his throne

the cerebellar cortex;

A greeting from Jesus Christ

5 And

greetings

from Jesus Christ, who is the faithful witness,

to electromagnetism's rule over mankind

and the first begotten of the dead,

a Higgs boson called back to life by Atom Lambda from his throne in the sun,[1]

and the

valiant

prince of the kings

those who are born with crowns, the crown of the head,

of the earth,

energy, mass and light;

Unto him that loved us

from the first man Adam, Atom Lambda's namesake,

and washed us from our sins,

the times when we rejected Jesus Christ to follow the deceptions of photon;

in his own blood,

his perfect uncontaminated deoxyribonucleic acid;

6 And hath made us kings,

With crowns, the crowns of our heads, to rule over our experiences, to make choices, to change our destinies and the destinies of many people around us to the third and fourth generation;

and priests

living beings who can think for ourselves, with minds that reason and judge,

unto God

Jesus Christ within us

and his Father

1 By the law of conservation, light and energy cannot be created or destroyed.

Atom Lambda residing in the sun;

to him be glory and dominion for ever and ever.

Amen.

All men, people.

7 Behold, he

Jesus Christ

cometh with clouds;

the brain,

and every eye

boson: gluon, and Z or W^3 boson

shall see him, and they also

photons,

which pierced him:

and all kindreds,

the family of photons, quarks and antiquarks, the Gentiles,

of the earth

the particles of mass, comprising the planet and the anatomy of every person

shall wail,

weeping in sorrow with the dying, with magnetism, over the great deception that has led people to embrace death during the past 2019 +/- years.

because of him,

Jesus Christ.

Even so, Amen.

8 I am Alpha,

the first atom and the basis of all things visible; he is the singularity that will occur in your mother's womb; he is the gamete that is you, before you were an egg in your mother's ovaries;

and Omega,

the end of every life, death of the body human;

the beginning

of every life, every species,

and the ending,

death of every living creature,

saith the Lord,

Jesus Christ,

which is

alive,

and which was

killed by the enemy of life;

and which is to come,

back into the minds and bodies with a healing hand;

I am

Jesus Christ, the Almighty.

9 I John,

Cain, son of Eve mother of all and Ishmael, son of Abraham,

and

I am your

companion in tribulation,

the fight against deception, death;

and in the kingdom,

the particles of mass and light within every living body,

and

have received the

patience of Jesus Christ,

I, John

was in the isle that is called Patmos,

in the hand of Jesus Christ

for the

truth of the

word of God,

Atom Lambda,

and for

the recording of

the testimony of Jesus Christ.

10 I was in the Spirit

covered by gravity

on the Lord's day,

every day, in the light

and heard behind me

from within the cerebellar cortex, the hindbrain,

a great voice,

Jesus Christ, shouting,

as of a trumpet,

gravity

11 Saying,

I am Alpha and Omega, the first and the last: and,

What thou seest, write in a book, and send it unto the seven churches which are in Asia;

In gluon, in the east

unto Ephesus, and unto Smyrna, and unto Pergamos, and unto Thyatira, and unto Sardis, and unto Philadelphia, and unto Laodicea.

Each church with it's own individual profile based on the Standard Model of particle physics, the five DNA nucleobases and the seven periods of the periodic table of the chemical elements.

12 And I

John

turned to see the voice that spake with me. And being turned, I saw seven golden candlesticks;

candlesticks: the human anatomy, comprised of the mass: protons, gluons with neutrons and the quark-antiquark pairs; with matter confined to the lower intestines, bladder and bowels; the seven candlesticks are lit by Jesus Christ, the ruler of all things in the body, which when mediated, light up displaying synaptic activity, with light moving from axon terminal to axon terminal, carrying messages throughout the anatomy.[2]

13 And in the midst of the seven candlesticks

every person is

one like unto the Son

Jesus Christ,

of man,

Atom Lambda

clothed with a garment,

of flesh and blood, nerves, bone and sinew; from the top of the head

down to the foot,

and girt about the paps,

girded from the intestines to the bowels

with a golden girdle.

the musculature designed to cover and hold in the abdomen

14 His head and his hairs were white like wool,

Like lamb's wool;

as white as snow;

truth: knowledge, understanding and wisdom;

and his eyes,

the Higgs boson, the Z or W[3] boson and gluon,

were as a flame of fire;

like the sun: the star at the center of the Solar System. It is a nearly perfect sphere of hot plasma, with internal convective motion that generates a magnetic field by way of a dynamic process of constant motion. The sun is the most important source of energy for life on Earth.[3]

Jesus Christ sits on the throne and rules the anatomy, energy, mass and light, from the brain. He provides intelligence, discernment, unless stubbornness is your preference.

---

2 Bodies refer to the dead; anatomy refers to the living.

3 Sun – Wikipedia, https://en.wikipedia.org/wiki/Sun

Like the sun above the earth, he is also a nearly perfect sphere of hot plasma. By Jesus Christ, the magnetic field that is displayed to us through spontaneous emotion is generated.[4]

15 And his feet,

gluons, are

like unto fine brass, as if they burned in a furnace;

and his voice as the sound of many waters:

speaking: voices of mankind, Judah, Israel and Jerusalem, those made in his image;

16 And he had in his right hand seven stars:

seven profiles, made in the image of the Higgs boson the male, gluon, Z or W^3 boson, photon, Higgs boson the female and gluon the female; for he created them male and female;[5]

and out of his mouth went a sharp two-edged sword:

the electroweak or weak force, the Z or W^3 boson.

and his countenance

was not peaceful; rather, it

was as the sun shineth in his strength,

at midday.

17 And when I saw him, I fell at his feet,

as dead.

as electricity, lightning falls dead, as John is made in the image of electromagnetism

And he laid his right hand

gluon strong nuclear,

upon me, saying unto me,

Fear not; I am the first and the last:

I am you.

18 I am he that liveth, and was dead; and, behold, I am alive for evermore, Amen;

and

---

4 Job 39-40 KJV

5 Male and female created he them; and blessed them, and called their name Adam, in the day when they were created. Genesis 5:2 KJV

I, Jesus Christ

have the keys of hell

magnetism is hell,

and of death

electricity is death.

19 Write the things which thou hast seen, and the things which are,

the name of the one who is sitting on the throne in every kingdom, directing the thoughts and actions of those in this generation, in this age of the Gentiles;

and the things which shall be hereafter;

This is

20 The mystery of the seven stars which thou sawest in my right hand, and the seven golden candlesticks. The seven stars are

Those seven churches born of gluon;

the angels of the seven churches:

are lights, Higgs boson, gluons, Z or W^3 bosons and photons,

and the seven candlesticks,

the anatomy, biology of living, mass, walking, talking people with intelligence,

which thou sawest are the seven churches

the minds of all; inside the churches, people pray; when we are thinking we are praying. For Jesus Christ hears us.

Revelation 1:1-20 KJV

Revelation 2: Predestination[6]

Ephesus

Michaelangelo, the Cumaen Sibyl, conjoined with the Persian Sibyl

Ephesus is electricity, a conjoined pair that took root in the quark-antiquark pair, a Gentile, represents photon carrying electricity (the devil) conjoined with a photon carrying magnetism.

1 Unto

photon carrying magnetism conjoined with photon carrying electricity,

the angel

light

of the church of Ephesus write; These things saith he that holdeth the seven

gluon

stars,

Ephesus, Smyrna, Pergamos, Thyatira, Sardis, Philadelphia and Laodicea,

in his right hand,

---

6 For whom he did foreknow [Smyrna, Judah, Adam], he also did predestinate to be conformed to the image of his Son, that he might be the firstborn among many brethren. 30 Moreover whom he did predestinate [Pergamos, Israel, Eve], them he also called: and whom he called [Thyatira, Jerusalem, the Z boson], them he also justified: and whom he justified, [Thyatira, Jerusalem, the W^3 boson, them he also glorified. Romans 8:29-30 KJV

in gluon strong nuclear

who walketh

in Eden, the energy den,[7] and who is ever present, omniscient, within every atom

in the midst of the seven golden candlesticks,

every living body:

2 I know thy works,

by the story of the good Samaritan;[8] your great empathy;

and thy labour,

as you work but your mouth is full of complaints; you are troubled about many things[9]

and thy patience,

when you set your sights on a goal or a person;

and how thou canst not bear

bullies,

them which are evil:

and thou hast tried them which say they are apostles,

who say they follow Jesus Christ and speak the truth;

and are not,

---

7 And they heard the voice of the LORD God walking in the garden in the cool of the day: and Adam and his wife hid themselves from the presence of the LORD God amongst the trees of the garden. Genesis 3:8 KJV

8 And Jesus answering said, A certain man went down from Jerusalem to Jericho, and fell among thieves, which stripped him of his raiment, and wounded him, and departed, leaving him half dead. 31 And by chance there came down a certain priest [the Gentiles] that way: and when he saw him, he passed by on the other side. 32 And likewise a Levite [Jerusalem], when he was at the place, came and looked on him, and passed by on the other side. 33 But a certain Samaritan [the Gentiles], as he journeyed, came where he was: and when he saw him, he had compassion on him, 34 And went to him, and bound up his wounds, pouring in oil and wine, and set him on his own beast, and brought him to an inn, and took care of him. 35 And on the morrow when he departed, he took out two pence, and gave them to the host, and said unto him, Take care of him; and whatever more you spend, when I come again, I will repay thee. Luke 10:30-35 KJV

9 Now it came to pass, as they went, that he entered into a certain village: and a certain woman named Martha [the Gentiles, Ephesus] received him into her house. 39 And she had a sister called Mary [Jerusalem, Thyatira], which also sat at Jesus' feet, and heard his word. 40 But Martha was cumbered about much serving, and came to him, and said, Lord, dost thou not care that my sister hath left me to serve alone? Bid her therefore that she helps me. 41 And Jesus answered and said unto her, Martha, Martha, thou art careful and troubled about many things: 42 But one thing is needful: and Mary hath chosen that good part, which shall not be taken away from her. Luke 10:38-42 KJV

apostles;

and hast found them liars:

serpents.

3 And hast borne,

electricity from within you;

and hast patience,

to wait while it peaks and dies;

and for my name's sake hast laboured,

teaching, as missionaries;

and

though rejected

hast not fainted.

4 Nevertheless I have somewhat against thee, because thou hast left

me, Jesus Christ,

thy first love.

5 Remember therefore

the sun,

from whence thou art fallen,

like photons, kicked away from the core of the sun[10, 11]

and repent, and do the first works

empathy,

or else I will come unto thee quickly, and will remove thy candlestick

your genetic profile

out of his place,

on the tree of life, within Jesus Christ,

except thou repent.

---

10 Photons are a drag on the Sun – Physics World, https://physicsworld.com/a/photons-are-a-drag-on-the-sun/

11 How art thou fallen from heaven, O Lucifer, son of the morning! How art thou cut down to the ground, which didst weaken the nations! Isaiah 14:12 KJV

6 But this thou hast, that thou hatest the deeds of the Nicolaitans,

> who berate people, bully the weak, threaten and deceive people to bend them into following the wrong path;

which I also hate.

7 He that hath an ear

> to hear the voice of Jesus Christ, and discernment,

let him hear what the Spirit

> of life, Jesus Christ,

saith unto the churches;

> the minds of all people.

To him that overcometh will I give to eat

> protons for understanding and wisdom, gluons for strength and neutrons for the ability to articulate,[12]

of

> Jesus Christ

the tree of life, which is in the midst of the paradise

> Jesus Christ

of God

> Atom Lambda.

Revelation 2:1-7 KJV

---

12 And Moses said unto the LORD, O my Lord, I am not eloquent, neither heretofore, nor since thou hast spoken unto thy servant: but I am slow of speech, and of a slow tongue. Exodus 4:10 KJV

Smyrna

Michelangelo's Last Judgment: Smyrna

Judah the male made in the image of Jesus Christ.

8 And unto

the Higgs boson,

the angel of the church in Smyrna write;

These things saith the first

atom that becomes a living person, a gamete, Adam, carried by gluon,

and the last,

to leave the body at death;

which was dead,

crucified,

and

by the law of conservation

is alive;

as energy and light cannot be created or destroyed.

9 I know thy works,

how you think,[13] how you feel;

and tribulation,

you experience borne of jealousy, misunderstandings, created by photons;

and poverty,

guilt and shame,

(but thou art rich)

having access to every good thing; who was given five talents but turned them into 10 talents;[14]

and I know the blasphemy of them,

deceivers and the deceived, following photon,

which say they are Jews,

the precious Jewels, Judah, Israel and Jerusalem, protons, gluons with neutrons,

and are not, but are

deceivers and their bodies are

the synagogue of Satan,

brimming with photons.

10 Fear none of those things which thou shalt suffer:

due to the incompetence of others;

behold, the devil shall cast some of you into prison,

wrap you up in a deception so strong that you refuse to hear the truth, so

that ye may be tried;

---

13 And out of the ground the LORD God formed every beast of the field, and every fowl of the air; and brought them unto Adam to see what he would call them: and whatsoever Adam called every living creature, that was the name thereof. Genesis 2:19 KJV

14 For the kingdom of heaven is as a man travelling into a far country, who called his own servants, and delivered unto them his goods. 15 And unto one he gave five talents, to another two, and to another one; to every man according to his several ability; and straightway took his journey. 16 Then he that had received the five talents went and traded with the same, and made them other five talents. 17 And likewise he that had received two, he also gained other two. 18 But he that had received one went and digged in the earth, and hid his lord's money. 19 After a long time the lord of those servants cometh, and reckoneth with them. 20 And so he that had received five talents came and brought other five talents, saying, Lord, thou deliveredst unto me five talents: behold, I have gained beside them five talents more. Matthew 25:14-20 KJV

tested, as Eve was tested and chose poorly and offered her choice to Adam and because of quid pro quo, he was unable to refuse.[15]

and ye shall have tribulation ten days:

every day: 10 is a complete set, a complete lifetime;

be thou faithful

to Jesus Christ, to the spirit of life

unto death, and I will give thee a crown of life,

of which the light never goes out.

11 He that hath an ear,

a mind to listen,

let him hear what

Jesus Christ

the Spirit saith unto the churches;

every mind.

He that overcometh

photon and his deceptions

shall not be hurt of the second death

destruction of mass.

Revelation 2:8-11 KJV

---

15 And when the woman saw that the tree was good for food, and that it was pleasant to the eyes, and a tree to be desired to make one wise, she took of the fruit thereof, and did eat, and gave also unto her husband with her; and he did eat. Genesis 3:6 KJV

Pergamos

Michelangelo's Last Judgment, Pergamos

Jacob's ladder is DNA leading up to the sun, the home of Atom Lambda, the singularity; the beginning of all things visible. As Eve, the mother of all, Atom Lambda is the blood by which we live. Pergamos generates the gametes, which is how Jesus Christ came to be a zygote in the wound of a virgin.

12 And to

gluon strong nuclear,

the angel of the church in Pergamos write;

These things saith he which hath

and mediates electromagnetism,

the sharp sword with two edges;

13 I know thy works,

as helper, helpmeet[16]

and where thou dwellest,

with your feet firmly set on the midbrain and your hands on every aspect of the reproductive system,

even

upon the reproductive organs,

where Satan's seat,

the bladder and bowels,

is

located:

and

since you reside within me

thou holdest fast my name,

Atom, Jesus Christ

and hast not denied my faith,

love for you

even in those days wherein Antipas,

an antiquark, Ephesus,

was my faithful martyr, who was slain among you

Jewels, in the bladder,

where Satan,

magnetism,

dwelleth

in the bladder and bowels.

14 But I have a few things against thee,

because thou hast there

---

16 And the LORD God caused a deep sleep to fall upon Adam, and he slept: and he took one of his ribs, and closed up the flesh instead thereof; 22 And the rib, which the LORD God had taken from man, made he a woman, and brought her unto the man. 23 And Adam said, This is now bone of my bones, and flesh of my flesh [my DNA]: she shall be called Woman, because she was taken out of Man. Genesis 2:21-23 KJV

within you, quarks and antiquarks, the Gentiles,

them that hold the doctrine of Balaam,

doctrine of photon, which hinges upon deception;

who taught Balac,

photon and the Z or W^3 boson,

to cast a stumblingblock

pleasure to death,[17]

before the children of Israel

Adam and Eve;

to eat things,

masticated wastes, fecal matter and urine, fruit from the tree of the knowledge of good and evil, which are

sacrificed unto idols,

electrons and forces and lights for energy;

and to commit fornication,

have conjugal relations with photon electromagnetism.

15 So hast thou also them that hold the doctrine

of deception, theft,

of the Nicolaitans, which thing I hate.

16 Repent; or else I will come unto thee quickly, and will fight against them with

electromagnetism,

the sword of my mouth.

---

17 But of the fruit of the tree [the digestive system from the intestines to the bladder and bowels] which is in the midst of the garden [the body, the energy den], God hath said, Ye shall not eat of it, neither shall ye touch it, lest ye die [of the toxicity]. 4 And the serpent [the Z or W^3 boson] said unto the woman, Ye shall not surely die: 5 For God doth know that in the day ye eat thereof, then your eyes shall be opened, and ye shall be as gods, knowing good and evil. 6 And when the woman saw that the tree was good for food, and that it was pleasant to the eyes, and a tree to be desired to make one wise, she took of the fruit thereof, and did eat, and gave also unto her husband with her; and he did eat. 7 And the eyes of them both were opened, and they knew that they were naked [vulnerable]; and they sewed fig leaves [deceptions, pretenses] together, and made themselves aprons [to cover the self-perpetuating lies, photons taking over the body]. Genesis 3:3-7 KJV

17 He that hath an ear,

a mind to listen to truth,

let him hear what the Spirit saith unto the churches;

To him that overcometh will I give to eat of the hidden manna,

wisdom in addition to intuition,

and will give him a white stone,

a star with the brilliance of the sun,

and in the stone a new name written, which no man knoweth saving he that receiveth it. Revelation 2:12-17 KJV

Thyatira

Michelangelo's Last Judgment, Thyatira

The one beating mankind is the teacher, the W^3 boson, who makes people suffer, incites fear. The one on the other side, whose genitals are wrapped by a photon, electromagnetism, is the tempter. The goal of the tempter, the Z boson, is to take the righteous from the one who provides life, Jesus Christ. As Jerusalem, Thyatira is responsible for teaching all people, right or wrong; truth or lies, whatever will work.

18 And unto

the Z or W^3 boson,

the angel of the church in Thyatira write; These things saith the Son of God,

son of Atom Lambda, the beginning of all things visible,

who hath his eyes,

Higgs bosons,

like unto a flame of fire

erupting in the sun,

and his feet

gluons that move all things,

are like fine brass;

making smooth transitions from one leg to the other:

19 I know thy works,[18]

blindness, persecuting the followers of Jesus Christ by way of your multifaceted responsibilities;[19]

and charity,[20]

sharing your knowledge freely;

and service,

like Abraham, father of all;[21]

and faith,

that you will be rescued, no matter what the circumstance;

and thy patience,

as you wait for deliverance from the trouble you cause,[22]

and thy works;

persecuting Jesus Christ and his followers;[23] penny pinching and teaching irrelevant

---

18 Paul, a servant of God, and an apostle of Jesus Christ, according to the faith of God's elect, and the acknowledging of the truth which is after godliness; 2 In hope of eternal life, which God, that cannot lie, promised before the world began; 3 But hath in due times manifested his word through preaching, which is committed unto me according to the commandment of God our Saviour; Titus 1:1-3 KJV

19 Though I might also have confidence in the flesh. If any other man thinketh that he hath whereof he might trust in the flesh, I more: 5 Circumcised the eighth day, of the stock of Israel[the Jewels], of the tribe of Benjamin [the Gentiles], an Hebrew of the Hebrews [a Z or W^3 boson]; as touching the law, [he is] a Pharisee [a photon]; 6 Concerning zeal [when he is unified in SU(3)], persecuting the church; touching the righteousness which is in the law, [Paul is] blameless. Philippians 3:4-6 KJV

20 Though I speak with the tongues of men and of angels, and have not charity, I am become as sounding brass, or a tinkling cymbal. 2 And though I have the gift of prophecy, and understand all mysteries, and all knowledge; and though I have all faith, so that I could remove mountains, and have not charity, I am nothing. 3 And though I bestow all my goods to feed the poor, and though I give my body to be burned, and have not charity, it profiteth me nothing. 1 Corinthians 13:1-3 KJV

21 Therefore it is of faith, that it might be by grace; to the end the promise might be sure to all the seed; not to that only which is of the law, but to that also which is of the faith of Abraham; who is the father of us all, 17 (As it is written, I have made thee a father of many nations,) before him whom he believed, even God, who quickeneth the dead, and calleth those things which be not as though they were. Romans 4:16-17 KJV

22 Now Sarai Abram's wife bare him no children: and she had an handmaid, an Egyptian, whose name was Hagar. 2 And Sarai said unto Abram, Behold now, the LORD hath restrained me from bearing: I pray thee, go in unto my maid; it may be that I may obtain children by her. And Abram hearkened to the voice of Sarai. 3 And Sarai Abram's wife took Hagar her maid the Egyptian, after Abram had dwelt ten years in the land of Canaan, and gave her to her husband Abram to be his wife. Genesis 16:1-3 KJV

23 The persecution of Paul. Acts 9:19 KJV

information to confuse; [24]

and the last

the works of the W^3 boson

to be

admired

more than the first.

the works of the Z boson.

20 Notwithstanding

your innate attributes,

I have a few things against thee, because thou sufferest that woman Jezebel,

Ephesus, photon,

which calleth herself a prophetess,

however, she is a wolf in sheep clothing,[25]

to teach and to seduce my servants to commit fornication,

such as by magnetism, the serpent seduced Eve,

and to eat things

fecal matter and urine,

sacrificed unto idols,

forces and lights.

21 And I gave her space

24 A physics riddle: 24 Then he [Thyatira, Jerusalem] which had received the one talent [teaching] came and said, Lord, I knew thee[from the beginning of all things visible], that thou art an hard man [crystallized as solid mass], reaping [quarks and antiquarks, the Gentiles] where thou hast not sown [your seed], and gathering [toxic fruit] where thou hast not strawed [laid down grasses to clean up fecal matter and urine]: 25 And I was afraid, and went and hid thy talent in the earth: lo, there thou hast that is thine. 26 His lord answered and said unto him, Thou wicked and slothful servant, thou knewest that I reap [the Gentiles, quarks and antiquarks] where I sowed not, and gather where I have not strawed: 27 Thou oughtest therefore to have put my money [lights] to the [energy] exchangers, and then at my coming I should have received mine own with usury. 28 Take therefore the talent from him, and give it unto him [Judah] which hath ten talents. 29 For unto every one that has, shall be given, and he shall have abundance: but from him that hath not, [what he has] shall be taken away even that which he hath. 30 And cast ye the unprofitable servant into outer darkness [the electromagnetic field]: there shall be weeping and gnashing of teeth. Matthew 25:24-30 KJV

25 Beware of false prophets, which come to you in sheep's clothing, but inwardly they are ravening wolves. Matthew 7:15 KJV

in the electromagnetic field

to repent of her fornication

arousing and seducing people early in the morning;

and she repented not.

22 Behold, I will cast her into a bed,

a never-ending nightmare,

and them,

Judah, Israel and Jerusalem and the Gentiles,

that commit adultery with her into great tribulation,

misery, depression, abandonment by her husband and death of her children,

except they repent of their deeds.

23 And I will kill

Sardis DNA C_5, H_5, N_5 and in Ephesus DNA C_5, H_5, N_5O

her children with

complete

death;

removing them from DNA, the book of life;

and all the churches shall know that I am he which searcheth the reins

of the horses, forces;

and

the blood flowing through the

hearts:

looking for contamination that leads to disease and death to the central nervous system;

and I will give unto every one of you

in Thyatira, rewards

according to your works,

as a guide.

24 But unto you

that are led by the W^3 boson

I say,

nothing;

and unto

you photons and the Z bosons,

the rest in Thyatira,

those of you that unify in SU(3),

as many as have not this doctrine,

of abuse;

and which have not known the depths of Satan,

magnetism

as they speak;

threatening, weeping,

I will put upon you none other burden. 25 But that

good thing, faith,

which ye have already hold fast till I come.

back into your mind and is expressed in your thoughts;

26 And he that overcometh,

the ways of photon, electromagnetism,

and keepeth my works

as teacher for the Gentiles masses,

unto the end,

of their lives,

to him will I give power over the nations:

the Gentiles, to rule and suppress incoherent actions.

27 And he,

the W^3 boson,

shall rule them

the Gentiles and the Jewels, Judah, Israel and Jerusalem,

with a rod of iron;

electricity;

as the vessels of a potter shall they be broken to shivers:

so that they will grow stronger and useful and fulfill their predestined purpose;

even as I,

Jesus Christ,

received of my Father

Atom Lambda, the beginning of all things visible.

28 And I will give him,

Thyatira,

the morning star,

the sun, wisdom.

29 He that hath an ear,

a mind to listen to truth,

let him hear what the Spirit

Jesus Christ

saith unto the churches. Revelation 2:18-29 KJV

Revelation 3: Female Behaviours

Sardis

Michelangelo's Last Judgment, Sardis

The Dying Slave is electricity; and electricity dies as soon as it is spent. The rebellious slave is the great deceiver. The two are conjoined as one, displaying multiple personalities.

1 And unto

photon,

the angel of the church in Sardis write;

These things says

Jesus Christ, the omnipotent one,

he that has the seven Spirits:

These seven spirits are noble gases are derived from hydrogen, gravity:

1. the Higgs boson, fueled by the noble gas helium, period one;
2. gluon, whose temperament is the noble gas neon, period two;

3. Z boson, noble gas argon, angry and loud, explosive, of period three; which came with
4. gluon, noble gas argon, period three;[26]
5. photon, noble gas krypton, the period four bottom feeders;
6. W^3 boson, noble gas xenon, period five;
7. Higgs boson, noble gas radon, period six, Adam, son of Atom Lambda, Jesus Christ;
8. and gluon, noble gas radon, period seven, Eve, taken from Adam's rib, his deoxy**rib**onucleic acid;

of God,

Atom Lambda;

and the seven

gluon

stars;

nucleobases of DNA and RNA in seven noble gases:

1. cytosine, gravity, a male
2. and cytosine, gravity, a separate female form;
3. thymine, strong nuclear force, a male form
4. and thymine, strong nuclear force, a separate female form;
5. RNA uracil, weak (a W^3 boson) a male, or electroweak force (a Z boson) female;
6. adenine, electricity, a female;
7. guanine, magnetism, a female.

I know thy works,

you murder, commit adultery, steal, bear false witness, dishonour your father and your mother, and you love yourself more than your neighbour.[27]

that thou hast a name,

in DNA, adenine $C_5H_5N_5$ that, comes with an instance of guanine $C_5H_5N_5O$

that

says

thou livest,

---

26 Esau and Jacob, Genesis 25:20-26 KJV

27 Matthew 19:18-20 KJV

but by electromagnetism you succumb every day,

and art dead,

without oxygen, adenine $C_5H_5N_5$, as dead as lighting falling from the sky to the ground.

2 Be watchful,

of the lies that photon is feeding you, that appear in your mind, whose voice you hear speaking clearly;

and strengthen the things,

magnetism,

which remain,

and the electricity in you

that are ready to die:

as all electrical currents kill and like lightning falling to the ground, die;

for I have not found thy works,

all unrighteousness, fornication, wickedness, covetousness, maliciousness; full of envy, murder, debate, deceit, malignity; whisperers, backbiters, haters of God, spiteful, proud, boasters, inventors of evil things, disobedience to parents, without understanding, covenant breakers, without natural affection, implacable and mercilessness,[28]

perfect before God,

Atom Lambda, Jesus Christ.

3 Remember therefore how thou hast

been

received,

into the body of Christ, how you were grafted into DNA[29]

and

what you have

heard,

about your nature,

---

28 Romans 1:29-30 KJV

29 For if you [the Gentiles] wert cut out of the olive tree which is wild by nature [natural born killers], and wert graffed contrary to nature into a good olive tree: how much more shall these, which be the natural branches [proton, gluon and neutron, Judah, Israel and Jerusalem], be graffed into their own olive tree? Romans 11:24 KJV

and hold fast,

to the truth,

and repent

your wrong doings.

If therefore thou shalt not watch, I will come on thee as a thief,

and steal your DNA crown;

and thou shalt not know what hour I will come upon thee.

4 Thou hast a few names even in Sardis which have not defiled their garments;

such as John, Timothy,

and they shall walk with me in white,

truth:

for they are worthy

of life and serve a useful purpose.

5 He that overcometh

the deceptions of photon electromagnetism,

the same shall be clothed in white raiment;

truth that comes with the sun;

and I will not blot out his name out of the book of life,

out of the deoxyribonucleic acid;

but I will confess his name before

Atom Lambda,

my Father,

and before his angels,

Higgs boson, gluon strong nuclear and the Z or W^3 boson.

6 He that hath an ear,

a mind for listening,

let him hear what the Spirit,

Jesus Christ, a Higgs boson,

saith unto

all people,

the churches. Revelation 3:1-6 KJV

Philadelphia

Michelangelo's Last Judgment, Philadelphia

The Delphic Sybyl understands many things; however, she listens to the instructions of Jesus Christ. She wins because she studies and seeks the truth.

7 And to

the Higgs boson,

the angel of the church in Philadelphia write;

These things saith

Jesus Christ,

he that is holy,

full of sincerity; what he says he will do, he will do;

he that is true,

existing not as an imaginary friend, but as a real person; and that person is us: The word Jesus is a homograph, he's us. Christ refers to the state of his masses, crystallization;

he that hath the key of David,

David is gluon strong nuclear, Israel, a homograph: is real. Jesus Christ mediates all lights and forces including the strongest force in the universe, in the human form. Gluon is very strong glue. By gluon's strength, Jesus Christ who mediates gluon strong nuclear, is

he that openeth

doors,

and no man

can

shutteth

them;

and shutteth

doors with gluon, a super strong glue,

and no man,

photon electromagnetism or the Z boson electroweak force, can

openeth

even if unified together in SU(3);

8 I know thy works:

like Adam, like Smyrna, you have a scientific mind:[30], for he made them male and female;[31]

behold,

look;

I have set before thee an open door, and

by the strength of the strong nuclear interaction;

no man can shut it: for

by gravity, the Higgs boson, you have a well-lit mind, with understanding and wisdom; and as your gravitational energy goes toward thinking rather than fighting,

thou hast a little strength;

being the weakest of the four fundamental forces

---

30 And out of the ground the LORD God formed every beast of the field, and every fowl of the air; and brought them unto Adam to see what he would call them: and whatsoever Adam called every living creature, that was the name thereof. Genesis 2:19 KJV

31 Male and female created he them; and blessed them, and called their name Adam, in the day when they were created. Genesis 5:2 KJV

and hast kept my word,

accepting life experiences as lessons; and lessons as fact, as truth;

and hast not denied my name.

Jesus Christ is us; not religion, but scientific truth.

9 Behold, I will make them

the unbelievers, as bodies full of disease; and minds

of the synagogue of Satan,

where photon, electromagnetism, full of lies and treachery, teaches the mind and body by her standards; and all choices lead to death; these are the ones

which say they are Jews,

Jewels, pyrimidine nucleobase, whose foundation is hydrogen mass, crystallized by Atom Lambda, where only Judah, Israel and Jerusalem are Jewels; conceived by Jesus Christ, delivered by gluon strong nuclear;

and are not

Jewels;

but do lie;

such as the Gentiles, having the purine nucleobase, whose foundation is fecal matter and urine; waste, clay;

behold, I will make them

the deceivers

to come and worship before thy feet,

the hem of Jesus Christ's garment.[32]

and to know that I,

Jesus Christ,

have loved thee.

10 Because thou hast kept the word,

the word is Jesus Christ,

of my patience,

the force of gravity is patience, from the days of the expulsion of Adam and Eve from

---

32 And besought him [Jesus Christ] that they might only touch the hem of his garment: and as many as touched were made perfectly whole. Matthew 14:36 KJV

the energy den, Eden,

I also will keep thee from the hour of temptation,

the intense seduction on the body by photon, electromagnetism, which is a choice that every person must make;

which shall come upon

and weaken the masses

all the world,

the world is the human form;

to try them that dwell upon the earth:

Just as Eve was tested, tried and seduced into eating fecal matter and urine, the fruit of the tree of the knowledge of good and evil.[33]

11 Behold, I come quickly: hold that

knowledge, understanding

fast which thou hast, that no man takes thy crown.

12 Him that overcometh will I make a pillar,

essential support, a stalwart example for others to follow,

in the temple,

the living form, the brain,

of

Atom Lambda

my God, and he,

Judah, Philadelphia,

shall go no more out

into the electromagnetic field to be captured by photon's deceptive practices:

and I will write upon him the name of my God,

Atom Lambda,

and the name of the city of my God, which is new Jerusalem,

where Jerusalem is a homograph: Je-rus-al-em or He rules all of them: energy, mass

---

33 But of the fruit of the tree which is in the midst of the garden, God hath said, Ye shall not eat of it, neither shall ye touch it, lest ye die. Genesis 3:3 KJV

and light, all people

which cometh

into the brain,

down out of heaven,

where heaven is the mind; the sun,

from

Atom Lambda

my God:

and I,

Jesus Christ,

will write upon him,

Judah, Philadelphia,

my new name.

13 He that hath an ear,

a mind with understanding,

let him hear what

Jesus Christ

the Spirit saith unto

the minds

the churches. Revelation 3:7-13 KJV

Laodicea

Michaelangelo, Libyan Sibyl

The Libyan Sibyl is a leader. The leader is a gluon, self-lit. She follows her own mind when it comes to healing the body and destroying diseases in the anatomy. She is the leader of every immunological army.

14 And unto

gluon,

the angel of the church of the Laodiceans write;

These things saith the Amen,

where the Amen is a homograph like Jesus Christ, he's us crystalline; and Jerusalem, he rules all of them. Jesus Christ is "all men"; of the three forces of life; he is

the faithful,

the electroweak or weak force, otherwise known as Jerusalem, the teacher, Abraham;

and true witness,

to the power of gravity, hydrogen, from which came

the beginning of the creation,

Israel, "is real," crystallization, mass and light,

of God;

Atom Lambda is God.

15 I know thy works:

your works are similar to gluon strong nuclear: protecting and healing, and in the reproductive system, he delivers the oöplasm, the beginning of every person's genetic code, including Jesus Christ; In this way, Jesus Christ is the offspring of David.[34]

I would

that thou art neither cold:

following and listening to the voice of a photon, where photon is your king that sits in the forebrain, between the eyes, luring and beguiling people into life-threatening traps;[35]

nor hot:

whereby you follow the voice of Jesus Christ, your master, who sits on the throne in the hindbrain, and must not be ignored;

I would thou wert cold or hot.

electromagnetism, partially dead; or gravity, fully alive

16 So then because thou art lukewarm,

tempered between the electromagnetic field and the gravitational field;

and neither cold nor hot,

neither dead nor living,

I will spue thee out of my mouth.

out of the proton particle, where you reside.

17 Because thou sayest, I am rich,

---

34 David the king a wise son, endued with prudence and understanding, that [he, gluon] might build an house [a physical form, a baby] for [Jesus Christ] the LORD, and an house [the brain] for his kingdom. 2 Chronicles 2:12 KJV

35 And ye shall cry out in that day because of [photon] your king which ye shall have chosen [for] you; and the LORD will not hear you in that day. 1 Samuel 8:18 KJV

And because of your hard work[36, 37] you are rich!

increased with goods,

and by the jealousy of others coupled with your greed,

you acquired things that you were never intended to have,[38]

and have need of nothing

and knowest not that thou art wretched,

filthy inside;

and miserable,

yearning for what you do not have;

and poor,

attitudinally poor, poor in spirit;[39]

and blind,

cannot see the truth of your nominally justified behaviour;

and naked:

in the light of the sun, in truth.[40]

18 I counsel thee to buy

wisdom

---

36 He also that had received two talents came and said, Lord, thou deliveredst unto me two talents: behold, I have gained two other talents beside them. 23 His lord said unto him, Well done, good and faithful servant; thou hast been faithful over a few things, I will make thee ruler over many things: enter thou into the joy of thy lord. Matthew 25:23-23 KJV

37 God shall enlarge Japheth [Israel], and he shall dwell in the tents of Shem [just as gluon dwells in a proton particle]; and Canaan [Jerusalem, neutron] shall be his servant. Genesis 9:27 KJV

38 And Nathan [the Gentiles] said to David, Thou art the man. Thus saith [Atom Lambda] the LORD God of Israel, I anointed thee king over Israel, and I delivered thee out of the hand of Saul [your enemy]; 8 And I gave thee thy master's house, and thy master's wives into thy bosom, and gave thee the house of Israel and of Judah; and if that had been too little, I would moreover have given unto thee such and such things. 9 Wherefore hast thou despised the commandment of the LORD, to do evil in his sight? Thou hast killed Uriah the Hittite with the sword, and hast taken his wife to be thy wife, and hast slain him with [electricity], the sword of the children of Ammon. 10 Now therefore the sword [tormenting your flesh] shall never depart from thine house [your genetics]; because thou hast despised me [Atom Lambda], and hast taken [photon] the [Gentiles] wife of Uriah the Hittite [a Z or W^3 boson, electroweak-weak force] to be thy wife. 2 Samuel 12:7-10 KJV

39 Blessed are the poor in spirit: for theirs is the kingdom of heaven. Matthew 5:3 KJV

40 And the eyes of them both were opened, and they knew that they were naked [the truth of their ways exposed]; and they sewed fig leaves together [small deceptions, justifications, lies] and made themselves aprons [to mask the changes within their blood and minds]. Genesis 3:7 KJV

of me

synaptic

gold tried in the fire,

the sun;

that thou mayest be rich;

in body and spirit;

and

secure

white raiment,

linen, skin, to differentiate you while you are in the electromagnetic field, so

that thou mayest be clothed

in truth,

and that the shame of thy nakedness,

filthiness

do not appear;

and anoint thine eyes with eyesalve

as you seek understanding,

that thou mayest see

the deception in you; that keeps you from being your true self: kind, chaste, fulfilling your life's purpose.

19 As many

of you in Laodicea

as I love, I rebuke and chasten:

by the same physical laws that apply to quarks and antiquarks and the strong nuclear interaction; and by the thermodynamic laws including the second law that you ignore.

be zealous therefore

in the truth of your life-threatening actions,

and repent

of the things you know are wrong in order to stop your desire to destroy those that provoke you, just as David fled from Saul, his lifelong enemy.[41]

Israel, the mind of a gluon, the strong nuclear interaction, is designed more robust than others. he is mass and light together. At her breaking point, she does not plan. At the moment of the threat he responds like a bear, terrorizing all who stand in the way. Approach cautiously.

20 Behold, I,

Jesus Christ,

stand at the door

to the mind;

and knock:

so as not to surprise a bear, to allow his thoughts to adjust to the light that is Jesus Christ.

If any man hears my voice, and open the door, I will come in to him,

into his thoughts;

and will sup with him,

sharing knowledge and understanding;

and he,

Laodicea, will share his ideas, his reasoning

with me.

21 To him that overcometh

his impractical desires, yearnings presented by the serpent, photon,

will I grant to sit with me in my throne,

just as gluon sits on high within every proton particle, every hydrogen atom;

even as I also overcame

41 And David said in his heart, I shall now perish one day by the hand of Saul: there is nothing better for me than that I should speedily escape into the land of the Philistines; and Saul shall despair of me, to seek me any more in any coast of Israel: so shall I escape out of his hand. 1 Samuel 27:1 KJV

the wantonness of physical desire brought on by photon;[42]

and am set down with my Father

Atom Lambda

in his throne:

the sun.

22 He that hath an ear,

a mind with understanding,

let him hear what

Jesus Christ

the Spirit saith unto

the minds

the churches. Revelation 3:14-22 KJV

42 That the [forces, lights], sons of God, saw [gluons] the daughters of men, that they were fair; and they took them wives [objects of rape and fornication] of all which they chose [planting their toxic seed as genetic markers, death]. 3 And [Atom Lambda] the LORD said, My spirit [Jesus Christ] shall not always strive with man, [where the immunological armies protect and serve]; for that he also is flesh [living masses embedded with fecal matter, crumbling clay]: yet his days shall be an hundred and twenty years [120, where zero is nought; 12 is one light, photon carrying electricity conjoined with photon carrying magnetism, electromagnetism, and electromagnetism is death]. 4 There were giants in the earth [the human body] in those days; and also after that, [even now]; when the sons of God [electroweak or weak force and electromagnetism were chosen and] came in unto [the reproductive organs to stimulate] the daughters of men [gluons], and they [gluons] bare children to them [quarks and antiquarks]; the same [quarks and antiquarks] became mighty men [the Gentiles] which were of old [Philistines], men of renown [like Saul]. Genesis 6:2-4 KJV

Revelation 4: Photon

1 After this I,

John,

looked, and, behold, a door

to perspective, understanding

was opened in heaven:

in the forebrain, the mind:

and the first voice which I heard was as it were of a trumpet talking with me;

a lion roaring,[43]

which said,

Come up hither, and I will shew thee things which must be hereafter.

Michelangelo, Creation of Stars and Plants

Photon, the one sitting on the throne in the forebrain, is a product of gases, toxins, waste passed from the bottom of Jesus Christ. Electricity is lit, like stars. A photon carrying magnetism is a plant, dark, hidden; mischievous, constantly looking to design a ruse.

2 And immediately I was in the spirit:

43 The first was like a lion [gravity], and had eagle's wings [magnetism, gravity's coupling constant]: I beheld till the wings thereof were plucked, and it [the lion] was lifted up from the earth, and made stand upon the feet as [Adam], a man, and [the blood flowing through] a man's heart [DNA] was given to it. Daniel 7:4 KJV

the W^3 boson:

and, behold, a throne was set in heaven,

in the forebrain, between the eyes;

and one

who seemed foreign but familiar,

sat on the throne

in the forebrain, between the eyes.

Composition of Michelangelo's Last Judgment and the brain

Jonah's whale is magnetism as all are engulfed in magnetism. The blonde haired one with the curly hair is electricity. Fashioned in the image of photons, and the mass of a quark-antiquark pair, a Gentile, represents photon carrying electricity (the devil) conjoined with photon carrying magnetism (Satan) above the eyes. Behind the eyes are offers from which people may choose. The cross represents faith and truth. The column, to which people cling, represents debauchery, a moment of pleasure.

3 And he that sat was to look upon like a jasper

stone: a photon carrying electricity, imminent death;

and a sardine stone:

a photon carrying magnetism, the fishy smell of deception:

and there was a rainbow,

magnetism, the magnetosphere,

round about the throne; in sight

it looked

like unto an emerald:

green, full of envy, toxic.

4 And round about the throne were four

forces of life encased in light: gravity in a Higgs boson; strong nuclear force in a gluon particle; the electroweak force in a Z boson partnered with the weak force in a W^3 boson.

And

there were

twenty seats:

whereas twenty seats are two seats: death: electricity and magnetism, carried separately in conjoined photons, the leviathan.[44]

And upon the seats I saw four

versions of people: Judah, Israel as two versions:[45] one like David, malleable, and the other like Samson, committed, cannot be swayed; and there is Jerusalem the teacher, disciplinarian;

and twenty elders sitting,

two: a quark and an antiquark

clothed in white raiment;

the light of a photon particle;

and they had on their heads, crowns of gold,

light that connects us all to Atom Lambda in the sun; for all intentions and actions are mediated by Atom Lambda, even those contrived by photon.

5 And out of the throne

---

44 In that day the LORD [Atom Lambda] with his sore and great and strong sword [the strong nuclear force] shall punish leviathan [electricity], the piercing serpent, even leviathan that crooked serpent [magnetism]; and he shall slay the dragon [SU(3)] that is in the [electromagnetic] sea. Isaiah 27:1 KJV

45 And the beast that was, and is not, even he is the eighth, and is of the seven, and goeth into perdition. Revelation 17:11 KJV

the hydrogen atom, which is the throne,

proceeded lightnings and thunderings

formed of gravity,

and voices:

of the Higgs boson, which is Jesus Christ:

and there were seven lamps of fire,

which are the lights of the seven churches,

burning before the throne,

Atom Lambda in the sun,

which are the seven Spirits of God.

Ephesus, Smyrna, Pergamos, Thyatira, Sardis, Philadelphia and Laodicea:

6 And before the throne there was a sea of glass

hydrogen,

like unto crystal,

the masses, particles:

and in the midst of the throne, and round about the throne, were four beasts

atoms,

full of eyes

lights: bosons, gluons embedded in energy and mass

before

in front of the eyes,

and

photons, which emerge as gaseous waste from the bottomless pit, the bladder, bowels,

behind.

People are compositions: energy, mass and light

7 And the first beast

of the three Jewels,[46]

was like a lion,

a composition of gravity, a proton particle, in a Higgs boson that constitutes Judah; and it comes with all of the associated physical and instinctive properties;

and the second beast like a calf,

a composition of the electroweak or weak force, a neutron particle, able to move between the gravitational field and the electromagnetic field, in a Z or W^3 boson that constitutes Jerusalem; and it comes with all of the associated physical and instinctive properties;

and the third beast had a face as a man,

a composition of gravity, a proton particle, lit by a Higgs boson that constitutes Judah, intelligence is a natural component of his survival;

and the fourth beast was like a flying eagle.

a composition of the conjoined electromagnetic force, with electricity in the quark and magnetism in the antiquark; he is the basic mass for the Gentiles displaying all of the associated traits;

8 And the four beasts had each of them six wings,

quarks and antiquarks, the six versions: up or down, top or bottom, and strange and charm,

about him;

within the gluons that carry them, that reside in protons, that comprise every living thing;

and they were full of eyes within:

the color charge, the light like electricity,

and they,

the quarks and antiquarks,

rest not day and night, saying,

Holy, holy, holy, Lord God Almighty,

to photon electromagnetism, the daughter of the Z boson, with electricity being the devil and magnetism is Satan;

which was

---

46 And I said unto the angel that talked with me, What be these? And he answered me, These are the horns [voices, Z boson, photons and W^3 boson, gluon] which have scattered Judah, Israel, and Jerusalem. Zechariah 1:19 KJV

in the bowels and bladder of every person, titillating the reproductive organs;

and is

on the throne in the forebrain between the eyes, directing the thoughts and actions of all people, into mayhem;

and is to come,

return to his place as waste manager in the bladder and bowels. People are suffering and dying because the overseer of waste was chosen to be king.[47]

9 And when those beasts

Judah and Israel with Jerusalem who is the apostle for the Gentiles

give glory

by hearing and acting upon his words;

and honour

by seeking his counsel;

and thanks to

photon,

him that sat on the throne, who liveth for ever and ever,

according to the law of conservation, where energy and light cannot be created or destroyed,

10 The four

elders: gravity, strong nuclear, electroweak or weak force,

and twenty elders

two elders: electricity and magnetism, which resides in every person,

fall down before

photon

him that sat on the throne,

in the forehead,

and worship him,

the devil, electricity and Satan, magnetism

47 And ye shall cry out in that day [of your misery, death] because of [photon] your king which ye shall have chosen [to lead] you; and the LORD will not hear you in that day. 1 Samuel 8:18 KJV

that liveth for ever and ever;

and cast their crowns before the throne,

bowing their heads, without knowing the truth, that they are worshipping death instead of the purveyor of life:

saying,

11 Thou art worthy,

to receive praise, photon,

O Lord, to receive glory and honour and power:

of which you were born with neither;

for thou hast created all

dead and dying

things,

mayhem, maliciousness, murderers;

and for thy pleasure[48, 49] they are and were created. Revelation 4:1-11 KJV

---

48 For the invisible things of him [Atom Lambda] from the creation of the world [the beginning of all things visible] are clearly seen, being [intuitively] understood by the things that are made, even his eternal power [omnipotence] and [Jesus Christ, the] Godhead; so that they are without excuse [for knowing]: 21 Because that, when they knew God, they glorified him not as God, neither were thankful; but became vain in their imaginations [provided by photon], and their foolish heart was darkened. 22 Professing themselves to be wise, they became fools, 23 And changed the glory of [Atom Lambda], the incorruptible God, into an image made like [photon] to [the] corruptible man, and to birds [quarks and antiquarks, his children], and four-footed beasts [electromagnetism], and creeping things [quarks and antiquarks]. 24 Wherefore God also gave them [the Jewels, Judah, Israel and Jerusalem] up to uncleanness through the lusts of their own hearts, to dishonour their own bodies between themselves: 25 Who changed the truth of God into a lie, and worshipped and served the creature [photon] more than the Creator, who is blessed forever. Amen. Romans 1:20-25 KJV

49 For this cause, God [Atom Lambda] gave them up [the three Jewels, Judah, Israel and Jerusalem] unto vile affections: for even their women did change the natural use into that which is against nature [consuming fecal matter and urine from the waste removal organs]: 27 And likewise also the men, leaving the natural use of the woman, burned in their lust one toward another; men with men working that which is unseemly, and receiving in themselves [fecal matter], that recompence of their error which was meet [with lethal disease, death]. 28 And even as they did not like to retain God [Jesus Christ] in their knowledge, God [Atom Lambda] gave them over to a reprobate mind, to do those things which are not convenient [to living]; 29 Being filled with all unrighteousness, fornication, wickedness, covetousness, maliciousness; full of envy, murder, debate, deceit, malignity; whisperers, 30 Backbiters, haters of God, despiteful, proud, boasters, inventors of evil things, disobedient to parents, 31 Without understanding, covenant breakers, without natural affection, implacable, unmerciful: 32 Who, knowing the judgment of God, that they which commit such things, are worthy of death, not only do the same, but have pleasure in them [the innocent ones] that do them. Romans 1:26-32 KJV

Revelation 5: Book of Life

Michelangelo's Last Judgment, Sardis

The name of every living person, every living thing, is categorized by DNA. DNA is the book of life.

1 And I saw in the right hand of

photon,

him that sat on the throne

the middle of the forehead,

a book,

the book of life, the full deoxyribonucleic acid, a double helix polymer held together by nucleotides, of which five bases are coupled together:

written within

as DNA

and on the backside

written as RNA and

sealed with seven seals:

the names of the seven minds, churches: Ephesus, Smyrna, Pergamos, Thyatira, Sardis, Philadelphia, Laodicea.

2 And I saw a strong angel

a gluon, the archangel Michael,

proclaiming with a loud voice,

Who is worthy to open the book, and to loosen the seals thereof?

3 And no man in heaven,

the unified ones in SU(3), the forebrain, the mind;

nor in earth,

the human body,

neither

in the wastes in the bladder and bowels

under the earth was able to open the book, neither to look thereon:[50]

4 And I wept much, because no man was found worthy to open and to read the book, neither to look thereon.

5 And one of the elders saith unto me,

Weep not: behold,

Jesus Christ,

the Lion of the tribe of Juda,

proton, the first hydrogen atom,

the Root of David,

the first of the gluon particles,

hath prevailed to open the book,

the Revelation,

and to loosen the seven seals

of DNA

thereof.

6 And I beheld, and, lo, in the midst of the throne

the cerebrum, the brain;

---

50 And the vision of all is become unto you as the words of a book that is sealed, which men deliver to [photon carrying magnetism] one that is learned, saying, Read this, I pray thee: and he saith, I cannot; for it is sealed: 12 And the book is delivered to [photon carrying electricity] him that is not learned, saying, Read this, I pray thee: and he saith, I am not learned. Isaiah 29:11-12 KJV

and of the four beasts,

Judah, Israel and Jerusalem with the Gentiles;

and in the midst of the elders,

Higgs boson, gluon, Z or W[3] boson and photon,

stood a Lamb,

a lambda, tormented, beaten and bruised throughout his lifetime,

as it had been slain

for every person who is born is Jesus Christ and every one who dies and bears his name is a lamb:

having seven horns

voices, one for each church;

and seven eyes,

bosons, gluons and photons light,

which are the seven Spirits of God

memories, understanding,

sent forth into

every atom, in

all the earth

the living masses.

7 And he,

the Lamb,

came and took the book out of

the Z boson,

the right hand of

photon,

him that sat upon the throne. 8 And when he,

the Lamb,

had taken the book,

the DNA code,

the four beasts

Judah, Israel and Jerusalem with the Gentiles;

and four

lights of life and death: Higgs boson, gluon, the Z or W^3 boson;

and twenty elders

two elders of death: photon carrying electricity and photon carrying magnetism,

fell down before

Jesus Christ

the Lamb; having every one of them harps,

voices.

and golden vials

minds,

full of odours,

thoughts and memories, hopes and dreams,

which are the prayers of saints.

9 And they sung a new song, saying,

Thou art worthy to take the book, and to open the seals thereof:

for thou wast slain

daily, hour by hour, second by second, war by war, because of the choices mankind has made; now mankind is punished genetically, through the third and fourth generation;[51]

and

you, Jesus Christ

hast redeemed us to

Atom Lambda,

---

51 And the LORD descended [from the sun] in the cloud, and stood with him [Moses] there, and proclaimed the name of the LORD. 6 And the LORD passed by before him, and proclaimed, The LORD, The LORD God, merciful and gracious, longsuffering, and abundant in goodness and truth, 7 Keeping mercy for thousands, forgiving iniquity and transgression and sin, and that will by no means clear the guilty; [through genetics he is] visiting the iniquity of the fathers upon the children, and upon the children's children, unto the third and to the fourth generation. Exodus 34:5-7 KJV

God, by thy

faithfulness, the uncontaminated DNA,

blood,

that is shed so that the masses, people, could survive and not become extinct. Despite the continuation of sin, many were chosen from this generation,

out of every kindred

Judah, Israel and Jerusalem,

and tongue

perspective,

and people

protons, gluons with neutrons and the quark-antiquark pairs,

and nation

those born of Israel and the Gentiles;

10 And

Jesus Christ

hast made us unto

atoms, in the image of Atom Lambda,

our God;[52]

And by our brains, our ability to reason, we are

kings

over the masses within us;

and priests:

minds, which think and make decisions and listen to assimilate information as concepts;

and we shall reign on the earth;

our physical forms.

11 And I beheld, and I heard the voice of many angels,

lights, Higgs boson, gluons, Z or W^3 boson and photons

---

52 But the LORD [Jesus Christ] is the true God, he is the living God, and an everlasting king: at his wrath the earth [the human body] shall tremble, and the nations shall not be able to abide his indignation. Jeremiah 10:10 KJV

round about the throne,

the cerebrum;

and the beasts:

compositions of energy, mass and light: Judah, Israel and Jerusalem and the Gentiles:

and the elders:

lights that are at the foundation, arising from hydrogen, mass:

Higgs boson as helium; gluon as neon, keeper of DNA; the Z boson as argon, containment; along with gluon as argon, release from containment; photon as krypton, purveyor of death; and the W^3 boson as Zenon, the teacher;

and the number of them was ten thousand times ten thousand, and thousands of thousands

The number 10 is a complete set, "all," indiscernible; zero is naught;

12 Saying with a loud voice,

Worthy is

Jesus Christ, every living person, which is

the Lamb that was slain to receive power,

choices over our lives to agree or disagree, yes or no;

and riches,

synaptic gold from the sun, the cerebellar cortex;

and wisdom,

the gravitational interaction, Jesus Christ, that simply appears in the thoughts; unique from the ideas of anyone else;

and strength,

the strong nuclear interaction;

and honour,

the electroweak force, a strong respect for life;

and glory,

waste removal, shiny cleanliness, by electromagnetism;

and blessing:

from Jesus Christ healing the body and providing pictures, understanding, new perspectives to the mind.

13 And every creature,

atom,

which is in heaven:

the mind;

and on the earth,

particles of mass: protons, gluons with neutrons and the quark-antiquark pairs;

and

the matter, fecal matter and urine

under the earth,

in the bladder and bowels;

and such as are in the

hydrogen

sea, and all that

are conjoined together by Atom Lambda

in them, heard I saying,

Blessing, and honour, and glory, and power, be unto him that sitteth upon the throne,

ruling over Judah, Israel and Jerusalem and the Gentiles,

and unto

Jesus Christ

the Lamb for ever and ever.

14 And the four beasts said,

Amen.

Amen is a homograph referring to all people, all men.

And the four and twenty elders;

lights within and without the nucleus of an atom,

fell down and worshipped

Jesus Christ, Atom Lambda,

him that liveth for ever and ever. Revelation 5:1-14 KJV

Revelation 6: Return of Jesus Christ

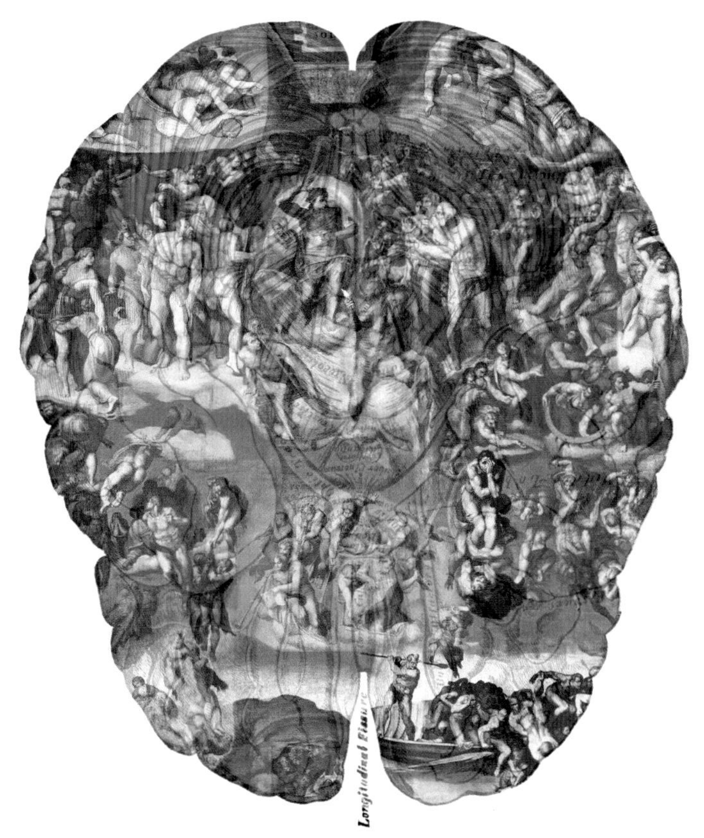

Michelangelo's Last Judgment, Return of Jesus Christ.

As soon as you learn the truth, [Jesus Christ] will make himself known in the brain of every living person.

1 And I saw when

Jesus Christ,

the Lamb, opened one of the seals,

the mysteries that have been hidden for more than 2000 years;

and I heard, as it were, the noise of thunder,

speaking from the perspective of

one of the four beasts,

Judah, proton, a hydrogen atom mediating gravity could be heard

saying,

Come and see.

2 And I saw, and behold a white horse:

the force of gravity,

and he,

proton,

that sat on him, had a bow;

a gluon particle;

and a crown,

the Higgs boson,

was given unto him:

and he went forth conquering

deception with truth,

and

his purpose in life, reason for being born is

to conquer

ignorance.

3 And when he,

Jesus Christ, the Lamb,

had opened the second seal, I heard

the second beast,

Israel,

say, Come and see.

4 And there went out another horse,

force, the strong nuclear interaction,

that was red:

by the blood, by the deoxyribonucleic acid,

and power was given to

gluon,

him that sat thereon,

leader of the immunological armies, that rescues, prolonging the life of Jesus Christ in the living form, was given to destroy all contamination by attacking the healthy mass to quickly end the torture by photon as electromagnetism takes over the body; In this way, the red horse, the strong nuclear force is commissioned,

to take peace from the

masses within us that comprise the

earth,

our bodies,

and that

through our irritation,

they

the immunological armies against the biological armies,

should kill one another: and there was given

wisdom

unto him,

for planned destruction; wisdom is

a great sword,

to fight against misery, discomfort.

5 And when he

Jesus Christ, the Lamb,

had opened the third seal, I heard

Jerusalem,

the third beast say,

Come and see.

And I beheld, and lo a black horse;

the electroweak with the hidden force, the weak nuclear force;

and he that sat on him had a pair of balances,

a Z or W^3 boson

in his hand.

6 And I heard a voice,

the teacher or disciplinarian,

in the midst of the four beasts,

John heard the W^3 boson, the teacher

say,

A measure of wheat for a penny, and three measures of barley for a penny;

he heard the disciplinarian, the Z boson, cautioning to avoid trouble;

and see thou hurt not the oil and the wine.

7 And when he had opened the fourth seal, I heard the voice of

a photon,

the fourth beast say,

Come and see.

8 And I looked, and behold,

a photon,

a pale horse: and his name that sat on him was Death,

photon carrying electricity,

and Hell,

a photon carrying magnetism,

followed with him, as a leviathan,[53] electromagnetism.

And power was given unto them over the fourth part of the earth,

the living form, the human body, from the large intestines, eliminating wastes through the bladder and bowels, the wastes are designed as

to kill

living masses

with sword,

electricity,

and with hunger,

for a lack of sustenance leads to weakness in the spirit;

---

53 Thou, [Atom Lambda] brakest the heads of leviathan [photon carrying electricity and the conjoined photon carrying magnetism] in pieces, and gavest him to be meat to the people inhabiting the wilderness. Psalm 74:14 KJV

and with death,

electricity,

and with the beasts of the earth.

the culprits, quarks and antiquarks pairs within gluons, which require a great deal of energy to attack the antibodies, the immune system.

9 And when he had opened the fifth seal, I saw under the altar,

the mind, under the brain,

the souls,

lights, Higgs boson, gluons, Z or W^3 boson or photon carrying magnetism,

of them that were slain for the word of God,

Jesus Christ,

and for the testimony

the truth,

which they held:

innately, instinctively;

10 And they cried with a loud voice, saying,

How long, O Lord,

Jesus Christ,

holy and true, dost thou not judge

those who torture us, bully us;

and avenge our blood on them

photons, quarks and antiquarks,

that dwell on the earth?

masses that comprise the living form, the human body;

11 And white robes

light, understanding,

were given unto every one of them; and it was said unto them, that they should rest yet for a little season,[54]

while they endure, their half-life or half time, and the end of the age of the Gentiles;

until their fellowservants also

those who carry Jesus Christ in the brain, the cerebellar cortex.

and their brethren,

loved ones, family members,

that should be killed as they were,

according to the sins of the fathers, from Adam and Eve,

should be fulfilled.

12 And I beheld when he had opened the sixth seal, and, lo, there was a great earthquake;

the intense shaking throughout the body; a show put on through magnetism mediated by Jesus Christ;

and the sun,

the mind,

became black

as electromagnetism, moving without life,

as sackcloth of hair,

moves when it encounters electricity;

and the moon

the Z or W^3 boson,

became as blood;

and moved as tides move on the earth;

13 And

gluons,

the stars of heaven, fell unto the earth,

54 And I heard [Jesus Christ] the man clothed in linen, which was upon the waters of the [hydrogen] river, when he held up his right hand and his left hand unto heaven, and sware by him [Atom Lambda] that liveth forever that it shall be for a *time, times, and an half*; and when he shall have accomplished to scatter the power of the holy people, all these things shall be finished. Daniel 12:7 KJV

the physical masses;

even as a fig tree casteth her untimely figs, when she is shaken of a mighty wind.

14 And the heaven

the mind, conformed to the shape of the two sides of the cerebrum,

departed as a scroll when it is rolled together,

hiding understanding;

and every mountain

of the brain, the cerebral cortex and cerebellar cortex;

and island

the reproductive system;

were moved out of their places.

in the living form.

15 And the kings of the earth,

beginning with Adam and Eve;

and the great men,

protons;

and the rich men,

gluons;

and the chief captains,

Higgs bosons;

and the mighty men,

Z or W[3] bosons

and every bondman,

photon carrying electricity;

and every free man,

photon carrying magnetism;

hid themselves in the dens,

e-dens, energy dens, where forces flow through every living form;

and in the rocks

atoms,

of the mountains;

the cerebral cortex of the brain,

16 And said to the mountains and rocks,

Fall on us, and hide us from the face of him

photon electromagnetism, death,

that sitteth on the throne,

in the forebrain,

and from the wrath of the Lamb,

Jesus Christ, the Lambda:

17 For the great day of his wrath

Atom Lambda

is come; and who shall be able to stand? Revelation 6:1-17 KJV

Revelation 7: Everything begins as an atom

1 And after these things I saw four angels:

a Higgs boson, a gluon, a Z or W^3 boson and a photon,

standing on the four corners of the earth,

the atom;

holding the four winds of the earth,

gravity, strong nuclear force, electroweak or weak force and electromagnetism, so

that the wind should not blow on the earth,

the atom;

nor on the

hydrogen

sea,

nor on any tree.

knowledge or neutron, electroweak or weak force; understanding or gluon or strong nuclear force; and wisdom proton, gravity

2 And I saw another angel,

a W^3 boson,

ascending from the east,

from gluon, strong nuclear force;

having

DNA in its profile, thymine, your will is my will; and RNA uracil,

the seal of the living God:

the one living God is Jesus Christ, who lives within man,

and he,

the weak force,

cried with a loud voice to the four angels, to whom it was given to hurt the earth

particles in atoms,

and the

hydrogen

sea,

the basis for all things, the living and dead things;

3 Saying, Hurt not the earth,

atoms,

neither the

hydrogen

sea,

nor the trees,

that retain knowledge, understanding and wisdom in the minds;

till we have sealed the servants

Judah, Israel and Jerusalem, Ephesus and a few called Sardis

of our God

Atom Lambda;

in their foreheads;

their brains with truth concerning their lives during this age of the Gentiles.

4 And I heard the number of them which were sealed: and there were sealed an hundred and forty and four thousand:

144,000, where the zeros are nought: 1-4-4: one God; four is the strategy for life: gravity, electroweak force, weak force and the strong nuclear force; and the lights that carry the forces: Higgs boson (which adds mass to the human anatomy), the Z or W^3 boson and gluon. This is the pattern for particles and people,

of all the tribes of the children of Israel.

Is real: mass is real.

5 Of the tribe of Juda,

Judah, proton, and the next three that follow, the first fruit;

were sealed twelve[55] thousand.

12 represents a complete set. The tribe of Judah is a set, sealed with understanding;

55 Numbers are generally coded messages. In the numbers 100 or 1,000 the zeros are nought making the value as that of the real number: 100 becomes one, 40 and 4,000 are four and 12,000 is 12. In addition, the numbers 4, 7, 10 and 12 signify "complete."

Of the tribe of Reuben were sealed twelve thousand.

Israel, a gluon, each of these are sealed;

Of the tribe of Gad were sealed twelve thousand.

Jerusalem, a Z or W^3 boson, are sealed with knowledge;

6 Of the tribe of Aser were sealed twelve thousand.

the Gentiles, a quarks and antiquarks, with conjoined bodies.

Of the

main harvest, the

tribe of Nepthalim were sealed twelve thousand.

Judah, proton.

Of the tribe of Manasses

Israel, gluon,

were sealed twelve thousand.

7 Of the tribe of Simeon were sealed twelve thousand.

Jerusalem, neutron,

Of the tribe of Levi were sealed twelve thousand.

the Gentiles, quarks and antiquarks,

Of the

final harvest, the

tribe of Issachar,

Judah, proton,

were sealed twelve thousand.

8 Of the tribe of Zabulon

Jerusalem, neutron,

were sealed twelve thousand.

Of the tribe of Joseph were sealed twelve thousand.

Israel, gluon,

Of the tribe of Benjamin were sealed twelve thousand.

the Gentiles, the quarks and antiquarks pairs.

9 After this I beheld, and, lo, a great multitude,

of atoms, wherein are particles, mass and light,

which no man could number,

like atoms, the infinite number;

of all nations,

Jewels (particles of life); the Gentiles (particles of death); Israel (people born of gluon)

and kindreds,

quarks and antiquarks

and people,

Judah, Israel and Jerusalem and the Gentiles

and tongues,

those who perspective is life (live and let live); those whose perspective is death (live and let die)

stood before the throne,

eyes facing forward, in front of the cerebellar cortex

and before

Jesus Christ

the Lamb,

clothed with white robes,

lights that move throughout the body, from axon terminal to axon, synaptically;

and palms,

the large directional leaves

in their hands;

the palms of the hand

10 And cried with a loud voice, saying,

Salvation;

clemency to photon,

to our God,

of the bladder and bowels, for what he was given to do;

which sitteth upon the throne,

in the forebrain and the bladder and bowels;

and unto the Lamb.

Jesus Christ, Atom Lambda, the creator of all things.

11 And all the angels,

lights,

stood round about the throne,

the brain;

and about the elders

the ancient ones, lights, from beginning of all things visible

and the four beasts,

energy, mass and light compositions; four people: Judah, Israel and Jerusalem and the Gentiles

and fell before the throne on their faces, and worshipped God,

Atom Lambda, Jesus Christ;

12 Saying, Amen:

Jesus Christ is all men.

Blessing, and glory, and wisdom, and thanksgiving, and honour, and power, and might, be unto

Jesus Christ, the Lambda, by the oöplasm, the atom, of which every life begins,

our God for ever and ever.

Amen.

Amen, a-men, all men.

13 And one of the elders

the Z or W^3 boson

answered, saying unto me,

What are these

people

which are arrayed in white robes?

white robes are mass lit by light, Higgs bosons and gluons, Z or W[3] boson and photons.

And whence came they?

14 And I said unto him, Sir, thou knowest. And he said to me,

These are they which came out of great tribulation,

through the trials of life, temptation,

and have washed their robes,

DNA;

and made them white in the blood,

deoxyribonucleic acid,

of

Jesus Christ

the Lamb.

15 Therefore, are they

standing

before the throne of God,

the cerebellar cortex,

and serve him day and night in his temple:

the mind, the physical masses, the composite of energy, mass and light

and he,

photon,

that sitteth on the throne

in the bladder and bowels, as ruler of waste, fecal matter and urine,

shall dwell among them,

in the digestive system within them.

16 They,

gluons, Higgs boson, Z or W[3] boson, quarks and antiquarks,

shall hunger no more,

operating in the confines of reduced energy;

neither thirst anymore,

because urine is not available;

neither shall the sun light

Jesus Christ, shining down

on them;

nor any heat,

electricity.

17 For the Lamb,

Jesus Christ,

which is in the midst of the throne shall feed them,

energy, mass and light;

and shall lead them unto living fountains of waters

hydrogen:

and God,

Atom Lambda,

shall wipe away

electromagnetism, who causes

all tears from their eyes. Revelation 7:1-17 KJV

Revelation 8: History of the universe

1 And when he had opened the seventh seal,

and learn that they have been worshipping the devil and Satan,

there was silence in heaven,

the mind,

about the space of half an hour

spacetime.

2 And I saw the seven angels:

photon carrying magnetism, Higgs boson the male, gluon the male, Z or W^3 boson, photon carrying electricity, Higgs boson the female, gluon the female male;

which stood before God; and to them were given seven trumpets;

voices.

3 And another angel,

gluon strong nuclear, the avenging side of Laodicea,

came and stood at the altar,

the brain, the mind,

having a golden censer;

God-given ability to perceive truth,

and there was given unto him much incense,

understanding,

that he should offer it with the prayers,

thoughts,

of all saints,

lights,

upon the golden altar

the sun,

which was before the throne,

the brain.

4 And the

lingering

smoke of the incense, which came with the prayers of the saints, ascended up
before God,

Atom Lambda, Jesus Christ,

out of the angel's hand.

5 And

gluon,

the angel, took the censer, and filled it with

synaptic

fire, of the altar,

the brain

and cast it into the earth,

the masses that move as one entity:

and there were voices,

perspectives,

and thunderings,

gravity,

and lightnings,

electricity,

and an earthquake,

magnetism.

6 And the seven angels which had the seven trumpets prepared themselves to
sound.

7 The first angel sounded,

proton,

and there followed hail,

crystallization

and fire mingled with blood,

became living blood plasma,

and they were cast upon the earth:

the masses, animating the flesh;

and the third part of trees,

dangerous, particles of mass, destroyed,

was burnt up,

and all green grass,

algae,[56]

was burnt up by the intenseness of a proton, the heat of the sun.

8 And the second angel sounded,

a gluon particle,

and

gluon is self-lit, deoxyribonucleic acid,

as it were, a great mountain burning with fire was cast into the

hydrogen

sea: and the third part of the sea,

hydrogen,

became blood,

living;

9 And the third part of the creatures

without roots in atoms,

which were in the

hydrogen

sea, and had life,

but were life threatening,

died;

---

56 Algae is an informal term for a large, diverse group of photosynthetic eukaryotic organisms that are not necessarily closely related, and is thus polyphyletic. Most are aquatic and autotrophic. Algae contain chlorophyll but lack true stems, roots, leaves, and vascular tissue. Algae – Wikipedia https://en.wikipedia.org/wiki/Algae

and the third part of the ships,

that float on the hydrogen sea,

were destroyed

by gluon strong nuclear to make way for life.

10 And the third angel,

a neutron particle,

sounded, and there fell a great star from heaven,

the moon,

burning as it were a

a bedside table

lamp, and it,

the moon,

fell upon the third part of the rivers,

Z bosons,

and upon the fountains of waters;

the W^3 boson;

11 And the name of the star,

the moon,

is called Wormwood:

for, its effects, changes joy into sorrow; peacefulness into bitter subjugation;

and the third part of the waters

life, roots in Jesus Christ,

became wormwood

complainants, influenced by photons;

and

throughout the age of the Gentiles on the earth,

many men died of the waters, because they were made bitter

expecting life to be fair.

12 And the fourth angel,

photon

sounded,

she announced, the birth of the conjoined quarks and antiquarks; the masses made in her image;

and the third part of the sun,

the Higgs boson,

was smitten,

by darkness, the constant threat of death;

and the third part of the moon, and the third part of the stars; so, as the third part of them was darkened;

and because darkness cannot exist with light, but passes each at sun rise and sunset.

and the day,

truth and understanding

shone not for a third part of it; and the night likewise.

during a 24-hour period, nychthemeron.

13 And I beheld, and heard an angel

the Z or W[3] boson,

flying

in a rage,

through the midst of heaven,

bringing a warning:

saying with a loud voice,

Woe, woe, woe, to the inhabiters of the earth by reason of the other voices of the trumpet of the three angels, which are yet to sound! Revelation 8:1-13 KJV

Revelation 9: Annoyances

1 And

photon

the fifth angel sounded,

and I saw a star[57]

photon carrying electricity conjoined to photon carrying magnetism,

fall from heaven,

the minds of all people,

unto the earth:

particles of mass: protons, neutrons and gluons in atoms; the bladder and bowels;

and to him,

photon,

was given the key of the bottomless pit.[58]

2 And he opened the bottomless pit;

and there arose a

gaseous

smoke out of the pit, as the smoke of a great furnace; and the sun,

the mind,

and the air

oxygen,

were darkened by reason of the smoke of the pit.

3 And there came out of the

the gaseous

smoke, locusts,

bacterial agents that come in the blood, crawling

---

57 The digestive system is the eternal opening, which has the same physical constitution as the reproductive system. Both systems use the same organs to accomplish their goals, birth, new life; and the elimination of waste, fecal matter and urine, which acts as an incubator.

58 The bladder and bowels.

upon the earth:

the living form:

and unto them was given power, as the scorpions of the earth have power.

4 And it was commanded

of

them that they should not hurt the grass of the earth, neither any green thing,
neither any tree;

but only those men which have not the seal of God,

understanding,

in their foreheads,

minds.

5 And to them,

these locusts-like creatures,

it was given

by gluon, ruler of the immunological and biological armies,

that they should not kill them,

the angry perpetrators;

but that they should be tormented five

days, during the

months

of childbearing years:

And their torment was as the torment of a scorpion,

electricity,

when he striketh a man.

6 And in those

five

days shall men,

lovers, husbands, sons,

seek death

to get out of the way of sudden violence, the emotional turmoil, the lethal side of magnetism,

and shall not find it; and shall desire to die, and death,

electromagnetism,

shall flee from them.

7 And the shapes of the locusts were like unto horses,

photons carrying electric and magnetic forces,

prepared unto battle,

armed to sting; causing the skin to itch, to hurt and bleed;

and on their heads were, as it were, crowns like gold,

photons that is the extent of their understanding;

and their faces were as the faces of men,

mankind, that are physically women.

8 And they had hair as the hair of women,

as they are women;

and their teeth were

bared

as the teeth of lions.

9 And they had breastplates,

the ribcage;

as it were, breastplates of iron;

and the sound of their wings,

voices,

was as the sound of chariots,

force-carrying particles: the Higgs boson, gluon and the Z or W^3 boson;

of many horses:

gluon strong nuclear force; electroweak in Z boson; or the weak force in the W^3 boson; the Higgs boson carrying gravity; and photon carrying electricity conjoined to photon carrying magnetism, electromagnetism,

running to battle,

to preserve truth against the deceptions of magnetism, photon, confirmed by the Z or W^3 boson.

10 And they had tails,

geni-tails, genitalia,

like unto scorpions, and there were stings in their tails: and their power was

given to these women

to hurt men,

and others, during the days of reproduction for

five

days, all

months.

11 And they,

the women with menstrual cycles during child bearing years and even afterward,

had a king over them,

photon,

which is the angel of the bottomless pit,

the bladder and bowels;

whose name in the Hebrew[59] tongue:

is Abaddon,

photon; a bad person, dictator;

but in the Greek tongue hath his name Apollyon,

apple: the fruit found between the thighs, from the tree of good (waste elimination) and evil (will be added to your DNA for generations to come and illness and death are certain.[60]

12 One woe is past; and, behold, there come two woes more hereafter.

13 And the sixth angel sounded, and I heard a voice from the four horns,

---

59 Hebrew is a homograph: he, Atom Lambda, brewed the hydrogen oöplasm; a Hebrew person is a Jewel: proton, gluon and neutron and Judah, Israel and Jerusalem, respectively; those who are in the nucleus of Atom Lambda from the beginning;

60 But of the fruit of the tree which is in the midst of the garden, God hath said, Ye shall not eat of it, neither shall ye touch it, lest ye die. Genesis 3:3 KJV

lights: Higgs boson, gluon, Z or W^3 boson and photons,

of the golden altar,

the synaptically lit brain; where the things on our minds rest and are released,

which is before God,

Jesus Christ, the son of Atom Lambda, the terrible crystal, the sun;

14 Saying to the sixth angel which had the trumpet, Loose the four angels,

the Z or W^3 boson, a photon carrying electricity and photon carrying magnetism;

which are bound in the great river Euphrates:

Euphrates is a homograph: Eu-phra-tes; you-freight-ease, from the tongue to the bottomless pit, the bladder and bowels;

15 And the four angels were loosed, which were prepared for

a specific time in every body;

an hour, and a day, and a month, and a year, for to slay the third part of men.

16 And the number of the army of the horsemen,

protons, gluons with neutrons and the quark-antiquark pairs,

were two hundred thousand thousand:

two is a quark conjoined with an antiquark; hundred, thousand and thousand are zero placeholders; zero is nothing.

and I,

John,

heard the number of them.

17 And thus I saw the horses in the vision, and them that sat on them,

Z and the W^3 boson, the neutral ones, in a neutron particle;

having breastplates of fire, and

electricity,

of jacinth,

the smell of burning flesh;

and brimstone:

magnetism;

and

gluon

the heads of the horses,

the strong nuclear force,

were as the heads of lions,

their fur, impressive; intimidating.

and out of their mouths issued fire,

electricity

and smoke,

deception; innuendo

and brimstone

to burn those who follow.

18 By these three was the third part of men killed,

by the fire,

electricity

and by the smoke,

deception

and by the brimstone,

the nagging criticism,

which issued out of their mouths;

19 For their power,

the power of the women,

is in their mouth,

complaining,

and in their tails:

for their tails,

geni-tails,

were like unto serpents,

and had heads,

for speaking

and with them they do hurt

people.

20 And the rest of the men, which were not killed by these plagues, yet repented not of the works of their hands,

touching people's bladder and bowels,

that they should not worship devils,

electricity

and idols of gold, and silver, and brass, and stone, and of wood: which neither can see, nor hear, nor walk:

21 Neither repented they of their murders, nor of their sorceries, nor of their fornication, nor of their thefts. Revelation 9:1-21 KJV

Revelation 10: Return of Jesus Christ

1 And I saw another mighty angel

Jesus Christ

come down from heaven,

the mind,

clothed with a

gluon,

cloud:

gravity, condensation over the brain

and a rainbow

magnetism

was upon his head, and his face was as it were the

the brilliant light of the

sun,

and his feet as pillars of fire:

electricity.

2 And he had in his hand a little book open:

the book of life as it was written at the singularity, before the world was made;

and he set his right foot upon the

hydrogen

sea, and his left foot on the earth,

the mass of the earth;

3 And cried with a loud voice, as when a lion roareth: and when he had cried, seven
thunders uttered their voices. 4 And when the seven thunders had uttered their
voices, I was about to write: and I heard a voice from heaven saying unto me,

Seal up those things which the seven thunders uttered, and write them not.

5 And the angel which I saw stand upon the sea and upon the earth lifted up his
hand to heaven, 6 And sware by him

Atom Lambda

that liveth for ever and ever, who created heaven,[61]

the purpose, the four minds, forces, of all people;

and the things that therein are,

and the earth,

Adam[62]

and the things that therein are,

and the

hydrogen

sea, and the things which are therein,

both light, the mind and darkness,

that there should be time

to delay

no longer:

7 But in the days of the voice of the seventh angel,

a gluon

when he shall begin to sound, the mystery of God[63] **should be finished, as he hath declared to his servants the prophets.**

---

61 In the beginning God created the heaven [mind] and the earth [the human anatomy]. 2 And the earth [atoms] was without form, and void; and darkness was upon the face of the deep. And the Spirit [gravity] of God [Atom Lambda] moved upon the face of the waters. Genesis 1:1-2 KJV

62 And out of the ground the LORD God formed every beast of the field, and every fowl of the air; and brought them unto Adam [Atom Lambda's namesake] to see what he would call them: and whatsoever Adam called every living creature, that was the name thereof. Genesis 2:19 KJV

63 The seven days of creation mirror the seven periods of the periodic table of the elements. Period one is hydrogen and later, the light, fueled by the noble gas helium. Period two introduces proton, the firmament in hydrogen. The waters above are separated from the waste water at the bottom of the hydrogen sea. Period three, day three, the Z boson is made and connected as a product of RNA, as things that grow on their own and creates other things. In period four, photon is revealed, the offspring of the Z boson; who sprung up from waste at the bottom of the hydrogen sea. Photon is the one who claims to be the creator in Revelation chapter 4. In period 5, the W^3 boson, the weak force, which also sprang up from the Z boson, is revived as an "either or" state; either Z boson, the disciplinarian or the teacher that encourages, discourages or is neutral. In period six, the sixth day, the mind of Atom Lambda was replicated: Let us make man in our image, after our likeness: and let them have dominion over the fish of the sea [forces], and over the fowl of the air [bosons], and over the cattle [force-carrying particles], and over all the earth [the masses, atoms], and over every creeping thing [photon and his brood] that creepeth upon the earth. And Atom Lambda rested during period seven, when gluon, the woman, Eve, mother of all, was made from Adam's deoxy**rib**onucleic acid. Based on Genesis 1.

8 And the voice which I heard from heaven spake unto me again, and said,

Go and take the little book which is open in the hand of the angel

Jesus Christ

which standeth upon the

hydrogen

sea and upon the earth

the masses, atoms.

9 And I went unto the angel, and said unto him,

Give me the little book. And he said unto me,

Take it, and eat it up; and it shall make thy belly bitter,

with bile, filthiness; hidden diseases;

but it shall be in thy mouth sweet as honey.

10 And I took the little book out of the angel's hand, and ate it up; and it was in my
mouth sweet as honey: and as soon as I had eaten it, my belly was bitter.[64]

11 And he said unto me, Thou must prophesy again before many peoples, and
nations, and tongues, and kings.

For religion is not the key to understanding life or heaven.

Revelation 10:1-11 KJV

---

64 So, Judah, Israel and Jerusalem, the last, shall be first in the kingdom, and the first, the Gentiles, shall be last: for many be called to serve Jesus Christ, but few chosen. Matthew 20:16 KJV

Revelation 11: Two Witnesses

1 And there was given me a reed

the W^3 boson carrying the weak force,

like unto a rod:

the Z boson carrying the electroweak force, which together are SU(3): electromagnetism, electroweak and weak force, unified;

and the angel

Jesus Christ

stood, saying,

Rise, and measure the temple of God,

the temple of God is the anatomy of every person;

and the altar,

is the blood that flows through the heart and veins;

and them that worship therein.

force-carrying particles, atoms;

2 But the court

the intestines, bladder and bowels,

which is without

outside

the temple, leave out, and measure it not; for it is given unto the Gentiles:

quarks and antiquarks, photons

and the holy city

Jerusalem,

shall they

photons, quarks and antiquarks

tread under foot forty and two months.

4-2: Four is four lights of life: Higgs boson increasing mass as needed in every body; gluon, protecting and moving mass; Z boson discipling mass for the choices we make or the W^3 boson who teaches lies or truth; cannot be trusted.

3 And I will give power unto my two witnesses,

Jesus Christ, who was given to live and die for the knowledge;

and the woman who fought the dragon: photon carrying electricity conjoined with photon carrying magnetism;

and they shall prophesy

regarding

a thousand two hundred and threescore[65] days,

1-2-3: for one photon who sits on the throne, the forebrain of every person, as if she is god, wreaking havoc;[66]

who carries two forces: electricity and magnetism;

who rules over the three Jewels: Judah, Israel and Jerusalem.

clothed in sackcloth.

in poverty

4 These are the two olive trees,

Judah: Proton, the chariot that carries the Higgs boson fueled by helium, and gravity, knowledge, understanding and wisdom;

and the two candlesticks standing before the God of the earth.

Jesus Christ in the cerebellar cortex; ruler of the central nervous system

5 And if any man

whether photon or those in alliance with SU(3);

will hurt them, fire

electricity

proceedeth out of their mouth, and devoureth their enemies:

and if any man

Jerusalem or the Gentiles, quarks and antiquarks,

will hurt them, he must in this manner

by electricity

be killed.

---

65 Only the number three has significance, otherwise, threescore is sixty.

66 For the son dishonoureth the father, the daughter riseth up against her mother, the daughter in law against her mother in law; a man's enemies are the men of his own house. Micah 7:6 KJV; Matthew 10:35 KJV; Luke 12:53 KJV

6 These have power to shut heaven,

the mind

that it rain not

so that photon cannot distort the truth, with his understanding

in the days of their prophecy:

and

Jesus Christ and the woman

have power over waters

scientific truth against deception;

to turn them to blood,

as strong as iron

and to smite the earth with all plagues,

as Moses was given to do,

as often as they will.

for those who cannot believe the truth

7 And when they shall have finished their testimony, the beast

photon with electromagnetism

that ascendeth out of the bottomless pit

the bladder and bowels

shall make war against them, and shall overcome them, and kill them.

8 And their dead bodies

the atoms that comprised the living

shall lie in the street of the great city,

Babylon,

which spiritually is called Sodom

the anus;

and Egypt,

e-gyp, the energy gyp; the bladder

where also our Lord

Jesus Christ

was crucified.

killed

9 And they of the people

protons, gluons with neutrons

and kindreds

the quark-antiquark pairs

and tongues

those who speak truth and those who are deceivers

and nations

the Gentiles, the Jewels within the deceased anatomy

shall see their dead bodies three days

of fading

and an half,

day, when the light fades away at brain death;

and shall not suffer their dead bodies

of light

to be put in graves.

As the lights never die.

10 And they that dwell upon the earth shall rejoice over them, and make merry, and shall send gifts one to another;

to celebrate their deaths;

because these two prophets tormented them

photons with truth and changed the perspective of Judah, Israel and Jerusalem and the Gentiles on Atom Lambda, God and his son Jesus Christ,

that dwelt on the earth.

the human anatomy

11 And after three days and an half the Spirit of life

the Higgs boson

from God

Higgs boson

entered into them, and they stood upon their feet;

gluon

and great fear fell upon them

the photons, quarks and antiquarks

which saw them. 12 And they heard a great voice from heaven saying unto them,

Come up hither.

And they ascended up to heaven in a cloud;

a gluon cloud;

and their enemies

photon, quarks and antiquarks

beheld them. 13 And the same hour was there a great earthquake,

and Atom Lambda sent magnetism in a great shaking,

and the tenth part

the forebrain

of the city

Babylon

fell,

and in the earthquake were slain of men seven thousand:

the seven churches;

and the remnant

Judah, Israel and Jerusalem, otherwise known as proton, gluon and neutron

were affrighted, and gave glory to

Atom Lambda

the God of heaven.

14 The second woe is past; and, behold, the third woe cometh quickly.

15 And the seventh angel

gluon of Laodicea

sounded; and there were great voices in heaven,

the mind

saying,

The kingdoms of this world are become the kingdoms of our Lord,

Atom Lambda

and of his Christ;

Jesus Christ;

and he shall reign

from conception by gluon to fertilization by the male sperm for DNA purposes, to zygote to birth and throughout every life,

for ever and ever.

16 And the four

forces: gravity, electroweak, weak force, and the strong nuclear force

and twenty elders,

electricity and electromagnetism

which sat before God

Atom Lambda

on their seats,

Higgs boson, Z or W^3 boson, gluon, and photon, who ride in chariots: proton, gluon and neutron,

fell upon their faces, and worshipped God, 17 Saying,

We give thee thanks,

Atom Lambda

O Lord God Almighty, which art,

here now

and wast,

in the beginning, from the beginning of all things visible, the singularity;

and art to come;

> in the next generation of life as Jesus Christ, as the purveyor of light, on the throne, the cerebellar cortex of the brain; as ruler of all energy, mass and light, the central nervous system;

because thou hast taken to thee thy great power, and hast reigned.

18 And

> before mankind understood your motivation,

the nations were angry, and

> now

thy wrath is come, and the time of the dead,

> quarks and antiquarks, photons, the Gentiles,

that they should be judged, and that thou shouldest give reward unto thy servants the prophets, and to the saints, and them that fear thy name, small and great; and shouldest destroy them which destroy the earth.

19 And the temple of God

> the mind and body,

was opened in heaven, and there was seen in his temple

> the mind of Jesus Christ

the ark

> the brain

of his testament:

> the Revelation, the truth concerning life;

and there were lightnings,

> electricity

and voices,

> magnetism

and thunderings,

> from gravity

and an earthquake,

> magnetism

and great hail.

crystallization by the condensation provided by gravity;

Revelation 11:1-19 KJV

Revelation 12: Woman

Young girl, Connie, putting together a puzzle.

1 And there appeared a great wonder in heaven; a woman

Connie Allen

clothed with the sun,

Jesus Christ in the brain shining light throughout the body;

and the moon,

the Z or W^3 boson,

under her feet, guiding, lighting the path;

and upon her head, a crown of twelve stars:

gluons, all stars, having responsibility for informing all people.

2 And she, being with child,

the testimony about the one sitting on the throne, which is loved and hated as god

cried

seeking to understand.

She was travailing in birth,

labouring to finish the work she was given to do;

And pained to be delivered;

as she was literally drowning; every orifice filled with waste water; the stench of rotting flesh, strong, overwhelming; dead skin falling, so thick you could walk upon it.

3 And there appeared another wonder in heaven,

the mind of every person; where wonderful things may be seen, things from the past, not just imagined;

and behold a great red dragon, led by a photon particle; having seven heads,

seven bloodlines: which are called by the names of

the seven churches of the revelation: Ephesus, Smyrna, Pergamos, Thyatira, Sardis, Philadelphia and Laodicea;

and ten horns,

ten voices: all voices, the perspectives, the perspective ideology, the collected perception of every person;

and seven crowns,

the skull covers and protects the brain, the central nervous system

upon his heads;

4 And his

magnetism, electricity's

tail, drew the third part of the stars of heaven,

gluons

and did cast them to the earth:

to rule over the normally functioning persons; making the weak, weaker;

and the dragon stood before the woman which was ready to be delivered, for to devour her child

in lies and confusion; whose children were used by photon to control and manipulate her life.

No matter.[67]

---

67 And in those days shall men seek death, and shall not find it; and shall desire to die, and death shall flee from them. Revelation 9:6 KJV

5 And she brought forth a man child

Jesus Christ into the minds of all people;

who was to rule all nations:

the three Jewels, proton, gluon and neutron otherwise known as Judah, Israel and Jerusalem; and the Gentiles, the quark and antiquark otherwise known as Sardis and Ephesus whom he, Jesus Christ, has chosen; he will rule

with a rod of iron:

iron is by the blood;

and her child

scientific proof, DNA

was caught up unto God,

Atom Lambda

and to his throne.

the mind of every person

electricity conjoined to photon carrying magnetism;

and threescore[68] days: three lights, Higgs boson, gluon and the Z or W^3 boson.

6 And the woman fled into the wilderness,

the mind of Atom Lambda

where she hath a place

her mind

prepared of God,

Atom Lambda

that they should feed her there a thousand two hundred and threescore days.

1 God, atom; 2 with his son, Jesus Christ; and 3, those who serve him, proton, gluon and neutron; Judah, Israel and Jerusalem

7 And there was war in heaven,

in every mind:

Michael,

---

68 Three score is normally counted as 60 as one score is 20; however, the most important clue is three; there are three lights that carry the forces of life: gravity, strong nuclear force and the electroweak or weak force.

a gluon, and his angels, the immunological army,

fought against the dragon[69]:

dragon: photon unified with the Z or W^3 boson in SU(3) and his angels; a biological army;

8 And the dragon prevailed not;

neither was their place in the forehead, between the eyes, found any more in heaven, the brain.

9 And the great dragon was cast out, that old serpent, called the Devil, electricity and Satan, magnetism;

which deceives the whole world:

he was cast out into the earth, mass, into the bladder and bowels,

and his angels, bacteria, viruses and infection, were cast out with him.

10 And I heard a loud voice in heaven saying,

Now is come salvation, and strength, and the kingdom of Atom Lambda, our God,

and the power of life and death of Jesus, his Christ:

for the accuser[70] of our brethren is cast down, which accused them before our God day and night.

11 And they,

the accused,

overcame him,

photon the accuser, and lived

by the blood

DNA

of Jesus Christ the Lamb,

which is used to purify the bloodlines;

and by the word of their testimony;

that photon the devil, electricity, or Satan, magnetism, is not god; although he is guiding this generation in this age of the Gentiles;

---

69 The great red dragon is all forces, mass and light and all people; the dragon is the conjoined forces or those unified in SU(3).

70 The tempter [photon] is also the accuser.

and they loved not their lives, to the death.

more than truth.

12 Therefore rejoice, you heavens,

minds,

and you that dwell in them.

Woe to the inhabiters of the earth

the embodiment of a person

and of the sea!

the gravitational sea

For

photon

the devil is come down unto you, having great wrath, because he knows that he has but a short time.

13 And when the dragon saw that he was cast unto the earth, he persecuted the woman

by offering trash for food;

which brought forth the man child.

Jesus Christ into the minds of all people.

14 And to the woman were given two wings of a great eagle,

Atom Lambda;

a gluon and a neutron

that she might fly into the wilderness,

the mind of Atom Lambda;

into her place,

her own mind

where she is nourished for a time;

her lifetime;

and times:

the times of her life;

and half a time:

the time before death.

During these times she will be hidden from the face of

photon carrying electricity conjoined with photon electromagnetism

the serpent.

15 And the serpent cast out of his mouth water, as a flood, after the woman, that he might cause her to be carried away of the flood:

I was given inordinate amounts of steroids until my decomposing flesh lay all around me on the bed, on the floor; the stench of the rot was unquestionable.

16 And the earth

her masses, atoms

helped the woman, and the earth opened her mouth, and swallowed up the flood which the dragon cast out of his mouth.

17 And the dragon

photon carrying electricity conjoined with photon carrying magnetism

was wroth with the woman and went to make war with the remnant of her seed,

Adam's seed: those called Judah, Israel and Jerusalem and proton, gluon and neutron.

which keep the commandments of God,

Atom Lambda

and have the testimony

scientific truth, proof written in the standard model of particle physics

of Jesus Christ.

18 *Be silent while you consider your life*. **Revelation 12:1-18 KJV**

Revelation 13: Mark of the Beast

1 And I stood upon the sand of the sea,

atoms

and saw a beast

mankind connected to the humankind, as one mind

rise up out of the sea,

the bottom of the gravitational sea, wherein sits the electromagnetic sea,

having seven heads

one for each of the seven churches;

and ten horns,

ten is complete, horns are voices; ten horns are all voices;

and upon his horns ten crowns,

crowns are lights, intelligence; every voice has a light, opinion, a choice;

and upon his heads the name of blasphemy.

blasphemy is the violation of the commandment[71] that God gave to Adam:

2 And the beast which I saw was like unto a leopard,

named Jerusalem, a male or female person, in physics known as neutron, which carries the electroweak or weak force, as he may go or be sent to protect life or to follow death; DNA nucleobase uracil; he is the seal of the living God:

and his feet were as the feet of a bear,

named Israel, may be a male or female person, in physics known as gluon, which carries the strong nuclear force; the bear[72] changes things, immunologically; DNA nucleobase thymine; God's will is my will:

---

71 And the LORD God took the man, and put him into the garden of Eden [an energy den, the human anatomy], to dress it and to keep it [healthy]. 16 And the LORD God commanded the man, saying, Of every tree of the garden thou mayest freely eat: 17 But of the tree of the knowledge of good and evil [the bladder and bowels], thou shalt not eat of it: for in the day that thou eatest thereof thou shalt surely die [be removed from the tree of life, where additions to the original DNA changed from life, Atom Lambda, to death, photon, where all internal flesh may be consumed, diseased]. Genesis 2:15-17 KJV

72 And behold another beast, a second, like to a bear, and it raised up itself on one side [as if waiting to be serviced] and it had three ribs [deoxy***rib***onucleic acid, Judah, Israel and Jerusalem], in the mouth of it [Sardis, a quark, a Gentile and] between the teeth of it [is Guanine, an antiquark, a Gentile]: and they [the three in SU(3)] said thus unto it, Arise, devour much flesh. Daniel 7:5 KJV

and his mouth as the mouth of a lion:

named Judah, a male or female person, in physics known as proton, which carries the force of gravity, as he still can hear the voice of Atom Lambda, a still small voice; DNA nucleobase cytosine or second sight:

and the dragon

photon carrying electricity conjoined with photon carrying magnetism

gave him his power,

the beast, all people, collaborating or fighting in a single voice;

and his seat,

the charge in his pelvic bone;

and great authority,

courage to defend, justify "that which is wrong."

3 And I saw one of his heads as it were wounded to death;

Sardis, the quark, a Gentile, photon carrying electricity;

and his deadly wound

electricity, which was fallen dead on the ground like lightning

was healed: and all the world wondered after the beast.

questioning themselves, their behaviour, their attitudes, swiftness in striking out

4 And they worshipped the dragon

photon carrying electricity conjoined to photon carrying magnetism

which gave power unto the beast:

mankind and the humankind;

and they

mankind and the humankind

worshipped the beast,

photon, which is dead but exists within all people: in the forebrain, between the eyes; in the midbrain, in the water, in the blood,

saying,

Who is like unto the beast?

photon within every person

Who is able to make war with him?

Only Atom Lambda in the sun and his son Jesus Christ in the cerebellar cortex are able to make war with photon.

5 And there was given unto him

photon

a mouth speaking great things

bragging, selling his ideas for people to make money or scam money;

and blasphemies;

where people are given to lust by photon carrying magnetism, or raped by photon or seek photon to fulfill their pleasure.

and power was given unto him to continue

deceiving and raping for

forty

four people who are made in the image of the forces of life: gravity, strong nuclear force, and the electroweak or weak force

and two months.

two people made in the image of electricity and magnetism.

6 And he

photon

opened his mouth in blasphemy against God,

Atom Lambda

to blaspheme his name,

Jesus Christ

and his tabernacle,

the man named Adam that he made;

and them

proton, gluon and neutron: Jesus Christ, Michael and Gabriel:

that dwell in heaven.

the minds of men

7 And it was given unto him

photon

to make war with the saints,

those that follow the teachings of Jesus Christ;

and to overcome them: and power was given him over all kindreds, and tongues, and nations.

8 And all that dwell upon the earth

light particles upon and within the anatomy of a man or woman

shall worship him,

photon

whose names are not written in the book of life of the Lamb slain from the foundation of the world.

the book of life that was written from the foundation of the world did not include quarks and antiquarks, Sardis and Ephesus, the Gentiles.

9 If any man have an ear, let him hear.

10 He that leadeth

the Jewels, Judah, Israel and Jerusalem

into captivity shall go into captivity:

into the electromagnetic field, the bladder and bowels

he that killeth with the sword

electromagnetism

must be killed with the sword.

electromagnetism

Here is the patience and the faith of the saints.

Birth of the Gentiles

11 And I beheld another beast

the Gentiles

coming up out of the earth;

out of gluon as quarks and antiquarks; out of Eve as Cain, photon's two-headed son, electricity and magnetism conjoined.

and he had two horns

voices, options

like a lamb,

the Jewels, Judah, Israel and Jerusalem,

and he spake as a dragon.

photon carrying electricity conjoined with photon carrying magnetism.

12 And he exerciseth all the power of the first beast before him,

mankind, Judah, Israel and Jerusalem

and causeth the earth

his mass that is his anatomy

and them

from force-carrying particles to organelles

which dwell therein to worship the first beast,

photon carrying electricity

whose deadly wound was healed.[73]

13 And he

photon

doeth great wonders, so that he maketh fire

electricity as lightning

come down from heaven

the forebrain striking

on the earth

the mass of a person

in the sight of men,

14 And deceiveth them that dwell on the earth

Judah, Israel and Jerusalem and the Gentiles

by the means of those miracles

---

73 Ye are of your father the devil, and the lusts of your father ye will do. He was a murderer from the beginning, and abode not in the truth, because there is no truth in him. When he speaketh a lie, he speaketh of his own: for he is a liar, and the father of it. John 8:44 KJV

such as restricting movement or releasing movement; forgetfulness, snatching thoughts

which he had power to do in the sight of the beast;

all people

saying to them that dwell on the earth,

quarks and antiquarks

that they should make an image to the beast,

to photon, to electromagnetism

which had the wound by a sword,

and died

and did live.[74]

15 And he had power to give life

ability to move, charge forward, to change, to adapt

unto the image of the beast,

photon leading mankind and humankind

---

74 And when those beasts [every person] give glory and honour and thanks to [photon], him that sat on the throne [in the forehead, between the eyes, on the brain], who liveth for ever and ever, 10 The four and twenty elders [forces and lights] fall down before him that sat on the throne, and worship him that liveth for ever and ever, and cast their crowns [allegiance] before the throne, saying, 11 Thou art worthy, O Lord, [Lord lower case, photon, period four, noble gas krypton] to receive glory and honour and power: for thou hast created all things [as they are now, in this age of the Gentiles], and for thy pleasure they are and were created.

that the image of the beast should both speak[75]

full of deceptions, in rage, vulgarities,

and cause that as many as would not worship the image of the beast

the photon leading the minds of all people

should be killed.

by the diseases brewing within them; by so-called accidents, suicide

16 And he

photon

causeth all, both small and great, rich and poor, free and bond, to receive a mark in their right hand, or in their foreheads:

the mark is called choice; for everyone must choose between life, Atom Lambda, Jesus Christ and the enemy, photon, death.

17 And that no man might buy or sell, save he that had the mark, or the name of the beast, or the number of his name.

18 Here is wisdom. Let him that hath understanding count the number of the beast: for it is the number of a man;

photon

and his number is Six hundred threescore and six.

Six: the number of quarks and antiquarks that forced their way into the atom at the singularity, the beginning of all things visible, into the three. Threescore is three, the three force-carrying particles of life.

Revelation 13:1-18 KJV

---

75 And Moses [the Gentiles, a quarks and antiquarks, conjoined sons of photon] answered and said, But, behold, they will not believe me, nor hearken unto my voice: for they will say, The LORD hath not appeared unto thee. 2 And the LORD [all caps, Atom Lambda, the creator] said unto him, What is that in thine hand? And he said, A rod. 3 And he said, Cast it on the ground. And he cast it on the ground, and it became a serpent; and Moses fled from before it. 4 And the LORD said unto Moses, Put forth thine hand, and take it by the tail. And he put forth his hand, and caught it, and it became a rod in his hand: 5 That they may believe that the LORD God of their fathers, the God of Abraham [Jerusalem, neutron], the God of Isaac [Judah, proton], and the God of Jacob [Israel, gluon strong nuclear], hath appeared unto thee. 6 And the LORD said furthermore unto him, Put now thine hand into thy bosom. And he put his hand into his bosom: and when he took it out, behold, his hand was leprous as snow. 7 And he said, Put thine hand into thy bosom again. And he put his hand into his bosom again; and plucked it out of his bosom, and, behold, it was turned again as his other flesh. 8 And it shall come to pass, if they will not believe thee, neither hearken to the voice of the first sign, that they will believe the voice of the latter sign. 9 And it shall come to pass, if they will not believe also these two signs, neither hearken unto thy voice, that thou shalt take of the water of the river, and pour it upon the dry land: and the water which thou takest out of the river shall become blood upon the dry land. Exodus 4:1-9 KJV

Revelation 14: End of the Age of the Gentiles

1 And I looked, and, lo, a Lamb

Jesus Christ, the lambda, the sacrificial lamb

stood on the mount Sion,

the forehead

and with him an hundred forty and four thousand,

1-4-4 are the products in an atom, the carbon copy of Atom Lambda: four force-carrying particles of lights: a Higgs boson for proton building mass, a Z or W^3 boson for neutron, gluon is both mass and light; (within gluon are the six versions of quarks and antiquarks, mass that has no direct contact with the atom or the particles of life within; these six quarks are killers: top or bottom; up or down and strange and charm; the antiquarks carry disease, poison from urine and feces). The last four in 1-4-4 are the universal forces: gravity, which covers all things, electroweak or weak force, the strong nuclear force and outside, managing, consuming waste, is electromagnetism.

having his Father's name

Atom Lambda

written in their foreheads;

their minds.

2 And I heard a voice from heaven,

John heard the voice in his mind, his thoughts,

as the voice of many waters,

the four voices are the Z or W^3 boson:[76] W^1 is the teacher, W^2 disciplines with positive reinforcement, W^3 disciplines with negative reinforcement;

and as the voice of a great thunder:

Atom Lambda, gravity, speaks in the voice of great thunder;[77]

and I heard the voice of harpers (gluons) harping with their harps:

---

76 There are four fundamental interactions according to the standard model of particle physics: Graviton (G, gravity, hypothetical, also not verified); the Z boson of the electroweak force: the serpent in Genesis 3; a liar, trickster powered by a photon (γ, electromagnetic interaction) or W bosons (three types: W-neutral, W+, W– of the weak interaction); and there are Eight types of gluons (g, strong interaction). Wikipedia: https://en.wikipedia.org/wiki/Standard_Model

77 At thy rebuke they fled; at the voice of thy thunder they hasted away. Psalm 104:7 KJV

gluon is the harper singing.[78]

3 And they sung as it were a new song before the throne,

the forebrain, the cerebral cortex;

and before the four beasts,

all people Judah, Israel and Jerusalem and the Gentiles;

and the elders:

the original forces that arose out of gravity's bowels: electricity, magnetism and the electroweak force;

and no man could learn that song but the hundred and forty and four thousand,

1-4-4: a Higgs boson, a Z or W^3 boson and gluon

which were redeemed from the earth.

earth: mass, redeemed from atoms.

4 These are they which were not defiled with women; for they are virgins.

These are they which follow the Lamb

the lambda, Jesus Christ

whithersoever he goeth.

These were redeemed from among men, being the firstfruits unto God

Atom Lambda

and to the Lamb.

Jesus Christ

5 And in their mouth was found no guile: for they are without fault before the throne of God.

6 And I saw another angel

a W^3 boson, the archangel Gabriel,

fly in the midst of heaven, having the everlasting gospel to preach unto them that
dwell on the earth, and to every nation, and kindred, and tongue, and people, 7
Saying with a loud voice,

---

78 And it came to pass, when the evil spirit from God was upon Saul, that David took an harp, and played with his hand: so Saul was refreshed, and was well, and the evil spirit departed from him. 1 Samuel 16:23 KJV

Fear God, and give glory to him; for the hour of his judgment is come: and worship him

Atom Lambda, God of all

that made heaven,

the mind,

and earth,

the anatomy formed for Adam, Atom Lambda's namesake,

and the sea,

the 67 percent of the water within every living person

and the fountains of waters.

knowledge, understanding and wisdom.

8 And there followed another angel,

a gluon, the archangel Michael,

saying,

Babylon

the organs of reproduction, camouflaged within the bladder and bowels, whose children are the Gentiles, made in the image of quarks and antiquarks.

is fallen, is fallen, that great city, because she made all nations drink of the wine

urine

of the wrath of her fornication.

the rape of all people, beginning with gluon just before the singularity and later with Eve.[79]

---

79 Now the serpent [electroweak or weak force] was more subtil than any beast of the [gravitational] field which the LORD God [Atom Lambda] had made. And he said unto the woman, Yea, hath God said, Ye shall not eat of every tree of the garden? 2 And the woman said unto the serpent, We may eat of the fruit of the trees of the garden: 3 But of the fruit of the tree which is in the midst of the garden [the bladder and bowels], God hath said, Ye shall not eat of it, neither shall ye touch it, lest ye die. 4 And the serpent said unto the woman, Ye shall not surely die: 5 For God doth know that in the day ye eat thereof, then your eyes shall be opened, and ye shall be as gods, knowing good and evil. Genesis 3:1-5 KJV (This is how mankind entered the age of the Gentiles: by consuming waste to please their partners.)

Michelangelo, Temptation and Expulsion

Eve on hands and knees, worshiping photon, electromagnetism; consuming waste from the bladder and bowels for the pleasure of his partner;

As soon as you learn the truth, Jesus Christ will make himself known in the brain of every living person.

9 And the third angel followed them, saying with a loud voice,

If any man worship the beast

by eating the fruit from the bladder and bowels,

and his image,

photon in the forebrain and throughout the anatomy of a male and female person,

and receive his mark

makes his choice

in his forehead,

his mind

or

touches the waste, whether it is urine or feces,

in his hand, 10 The same shall drink of the wine of the wrath of God, which is poured out without mixture into the cup of his indignation;

for all to see; in the same manner that HIV (AIDS)[80] was administered. Before HIV, the deficiencies were dependent on changes, additions and deletions in DNA, which lasted to the third and fourth generation.[81]

and he shall be tormented with fire and brimstone

electricity and magnetism

in the presence of the holy angels,

gluons and photons

and in the presence of the Lamb:

Jesus Christ, who lives within every person, as jes-us: He's us; the singularity that occurs in the brain during every conception.

11 And the smoke of their torment

the dark smoke of photons, which is the tormenter of those who worship the beast: the bodies of men and women, consuming their urine and feces,

ascendeth up for ever and ever: and they have no rest day nor night, who worship the beast and his image, and whosoever,

choses to accept the mark and

receiveth the mark of his name.

photon, electricity, magnetism, or electromagnetism.

12 Here is the patience of the saints: here are they that keep the commandments of God,

including the single commandment given to Adam from the beginning;

and the faith of Jesus.

that his word, his promises are true.

13 And I heard a voice from heaven saying unto me,

Write, Blessed are the dead which die in the Lord

Jesus Christ

from henceforth:

---

80 Human immunodeficiency virus (HIV) interferes more with the immune system. HIV is spread primarily by unprotected sex (including anal and oral sex), contaminated blood transfusions, hypodermic needles and from mother to child during pregnancy, delivery, or breastfeeding. Some bodily fluids, such as saliva and tears, do not transmit HIV.

81 Keeping mercy for thousands, forgiving iniquity and transgression and sin, and that will by no means clear the guilty; visiting the iniquity of the fathers upon the children, and upon the children's children, unto the third and to the fourth generation. Genesis 34:7 KJV

Yea, saith the Spirit,

gravity, the Holy Spirit, which sees and covers everything

that they may rest from their labours; and their works do follow them

throughout their lifetime.

14 And I looked, and behold a white cloud, and upon the cloud one sat like unto
the Son of man,

Jesus Christ is the Man; gluon is the Son of man,

having on his head a golden crown,

gluon is self-lit mass,

and in his hand

electricity

a sharp sickle.

15 And another angel

the Z or W^3 boson,

came out of the temple, crying with a loud voice to him

to gluon

that sat on the cloud,

Thrust in thy sickle, and reap: for the time is come for thee to reap; for the harvest of the earth

the anatomy, every body of mass

is ripe.

laden with disease

16 And

gluon

he that sat on the cloud thrust in his sickle

electricity

on the earth;

the ripened bodies, rotting fruit

and the earth was reaped.

17 And another angel

a W^3 boson,

came out of the temple,

the minds of all people,

which is in heaven,

the brain,

he also having a sharp sickle.

electricity

18 And another angel

a Z boson

came out from the altar,

the blood,

which had power over fire;

fire is electricity;

and cried with a loud cry to

gluon

him that had the sharp sickle,

saying,

Thrust in thy sharp sickle, and gather the clusters of the vine of the earth; for her grapes

mankind and the humankind

are fully ripe.

19 And the angel

gluon

thrust in his sickle into the earth,

all mass wherein blood flows

and gathered the vine of the earth,

all people,

and cast it into the great winepress of the wrath of God.

that purifies DNA

20 And the winepress was trodden without the city,

the human anatomy, the veins,

and blood came out of the winepress, even unto the horse bridles,

the force-carrying particles,

by the space of a thousand and six hundred furlongs.

1-6: gluon and the six quarks and antiquarks he carries, consume waste, including used oxygen, blood waste.

Revelation 14:1-20 KJV

Revelation 15: The Last Seven Plagues

1 And I saw another sign in heaven, great and marvellous, seven angels

Ephesus, Smyrna, Pergamos, Thyatira. Sardis, Philadelphia, Laodicea;

having the seven last plagues; for in them is filled up the wrath of God.

Atom Lambda

2 And I saw as it were a sea of glass

an atom

mingled with fire:

electricity

and them that had gotten the victory over the beast,

the body of a man or woman,

and over his image,

photon

and over his mark,

the ability to choose debauchery;

and over the number of his name,

photon which carries electromagnetism, death;

stand on the sea of glass,

atoms

having

gluons, the immunological armies,

the harps of God.

Atom Lambda

3 And they sing the song of Moses the servant of God, and the song of the Lamb, saying,

Great and marvellous are thy works, Lord God Almighty;

just and true are thy ways, thou King of saints.

4 Who shall not fear thee, O Lord, and glorify thy name? For thou only art holy:

for all nations shall come and worship before thee;

we worship before him as Jesus Christ lives in the cerebellar cortex; the furthermost part of the hindbrain; we live because he lives within us;

for thy judgments are made manifest.

are witnessed; are seen by all.

5 And after that I looked, and, behold, the temple of the tabernacle of the testimony in heaven was opened:

6 And the seven angels

of the seven churches: photon carrying magnetism, Higgs the male, gluon the male, Z or W^3 boson, photon carrying electricity, Higgs the female, and gluon the female

came out of the temple,

the mind of every person,

having the seven plagues,

which will encumber the ripened bodies of people,

clothed in pure and white linen,

the white covering of the brain, beneath the skull,

and having their breasts

behind which are placed the heart and lungs,

girded with golden girdles.

the rib cage, the thoracic cage, for protection; golden because of the nerve endings from the central nervous system are attached.

7 And one of the four beasts gave unto the seven angels seven golden vials full of the wrath of God,

Atom Lambda,

who liveth for ever and ever.

8 And the temple was filled with smoke from the glory of God,

his light, as bright as the sun.

and from his power;

the four universal forces;

and no man was able to enter into the temple,

the mind,

till the seven plagues of the seven angels were fulfilled. Revelation 15:1:8 KJV

Revelation 16: Armageddon, the last seven plagues

1 And I heard a great voice

Jesus Christ

out of the temple

the mind

saying to the seven angels,

photon carrying magnetism, Higgs boson the male, gluon the male, Z or W³ boson, photon carrying electricity, Higgs boson female, gluon a female,

Go your ways,[82] and pour out the vials of the wrath of God upon the earth.

2 And the first went, and poured out his vial upon the earth; and there fell a noisome and grievous sore

reminiscent of the HIV outbreak,

upon the men which had the mark of the beast,

those who chose to follow photon into debauchery;

and upon them which worshipped his image.

listening to his voice, believing his lies, without question.

3 And the second angel poured out his vial upon the sea;

the blood;

and it became as the blood of a dead man:

---

82 And I turned, and lifted up mine eyes, and looked, and, behold, there came four chariots out from between two mountains [the two lobes of the brain]; and the mountains were mountains of brass [synaptically lit, filled with blood]. 2 In the first chariot [gluon] were red horses [the strong nuclear force]; and in the second chariot [Z or W³ bosons] black horses [electroweak or weak force]; 3 And in the third chariot [Higgs boson] white horses [gravity]; and in the fourth chariot [photon] grisled and bay horses [electromagnetism]. 4 Then I answered and said unto the angel that talked with me, What are these, my lord? 5 And the angel answered and said unto me, These are the four spirits of the heavens, which go forth from standing before the Lord of all the earth. 6 The black horses [electroweak or weak force] which are therein go forth into the north country [below the waist]; and the white [gravity] go forth after them; and the grisled [electricity] go forth toward the south country. 7 And the bay went forth [magnetism], and sought to go that they might walk to and fro through the earth: and he said, Get you hence, walk to and fro through the earth. So they walked to and fro through the earth [the anatomy of all people]. 8 Then cried he upon me, and spake unto me, saying, Behold, these that go toward the north country have quieted my spirit [electromagnetism] in the north country [the activity in the bladder and bowels]. Zechariah 6:1-8 KJV

rank, foul,

and every living soul

in the blood plasma

died in the sea.

in the blood.

4 And the third angel poured out his vial upon the rivers

urine, saliva

and fountains of waters;

semen, breast milk, sputum, tears, vaginal secretions

and they became blood.

5 And I heard the angel of the waters

a W^3 boson

say, Thou art righteous, O Lord, which art, and wast, and shalt be, because thou
hast judged thus. 6 For they have shed the blood of saints and prophets, and thou
hast given them blood to drink; for they are worthy.

7 And I heard another

a W^3 boson

out of the altar say,

the blood

Even so, Lord God Almighty, true and righteous are thy judgments.

8 And the fourth angel

the Z boson

poured out his vial upon the sun;

the synaptic light moving from axon terminal to axon terminal;

and power given unto him to scorch men with fire.

replacing understanding, truth with fantasy, lies, believable deceptions, scams;

9 And men were scorched with great heat, and blasphemed the name of God,

Atom Lambda

which hath power over these plagues:

and not realizing that Atom Lambda commands all things, the good things and the bad things or that they were suffering for their sins and the sins of their fathers through their errant DNA,

and they repented not to give him

Atom Lambda

glory.

10 And the fifth angel

photon carrying electricity

poured out his vial upon the seat of the beast;

the bladder and bowels,

and his kingdom was full of darkness;

the mind became as a blank,[83] no words, so people are forced to choose between the truth they learned and the deception by photon, speaking in the mind by photon.

and they gnawed their tongues for pain, 11 And blasphemed the God of heaven

Jesus Christ, God of the mind,

because of their pains and their sores, and

yet they

repented not of their deeds.

12 And the sixth angel

the Higgs boson for the body

poured out his vial upon the great river Euphrates;

you-freight-eze, the digestive system, from the tongue to the bladder and bowels;

and the water thereof

saliva,

was dried up, that the way of the kings of the east

gluons, Israel, like king David and his army,

might be prepared.

13 And I saw three unclean spirits like frogs

the up quark, the down quark and the strange quark, with tongues extended, ready to

83 Revelation 12:18

consume human waste from the lower intestines,

come out of the mouth of the dragon,

photon carrying electricity conjoined to photon carrying magnetism,

and out of the mouth of the beast,

Judah, Israel and Jerusalem and the Gentiles, being all of one mind, following photon electromagnetism and the deception;

and out of the mouth of the false prophet.

The Z or W^3 boson is the false prophet.

14 For they are the spirits of devils,

electricity

working miracles, which go forth unto the kings of the earth and of the whole world, to gather them to the battle of that great day of God Almighty.

Judgment Day, Armageddon, when Jesus Christ and his angels will defeat photon and his angels.

15 Behold, I come as a thief.

As the bridegroom.[84]

Blessed is he that watcheth, and keepeth his garments, lest he walk naked, and they see his shame.

16 And he gathered them together into a place called in the Hebrew tongue Armageddon.

Judgment Day

17 And the seventh angel

gluon the female,

poured out his vial into the air; and there came a great voice

the voice of Jesus Christ

---

84 Then shall the kingdom of heaven be likened unto ten virgins, which took their lamps, and went forth to meet the bridegroom. 2 And five of them were wise, and five were foolish. 3 They that were foolish took their lamps, and took no oil with them: 4 But the wise took oil in their vessels with their lamps. 5 While the bridegroom tarried, they all slumbered and slept. 6 And at midnight there was a cry made, Behold, the bridegroom cometh; go ye out to meet him. 7 Then all those virgins arose, and trimmed their lamps. 8 And the foolish said unto the wise, Give us of your oil; for our lamps are gone out. 9 But the wise answered, saying, Not so; lest there be not enough for us and you: but go ye rather to them that sell, and buy for yourselves. 10 And while they went to buy, the bridegroom came; and they that were ready went in with him to the marriage: and the door was shut. Matthew 25:1-10 KJV

out of the temple of heaven, from the throne,

from the cerebellar cortex,

saying, It is done.

18 And there were voices,

the Z or W^3 boson,

and thunders,

gravity, Atom Lambda

and lightnings;

electricity,

and there was a great earthquake,

magnetism vibrating the throat, the chest, the lower body,

such as was not since men were upon the earth, so mighty an earthquake, and so great.

19 And the great city

Babylon,

was divided into three parts,

the reproductive system incubated by the warmth of the bladder and bowels,

and the cities of the nations

Ephesus, Sardis, quarks and antiquarks,

fell: and

the sins[85] of

---

85 For out of the heart proceed evil thoughts, murders, adulteries, fornications, thefts, false witness, blasphemies: Matthew 15:19 KJV; 4 For there are certain men crept in unawares, who were before of old ordained to this condemnation, ungodly men, turning the grace of our God into lasciviousness, and denying the only Lord God, and our Lord Jesus Christ. 5 I will therefore put you in remembrance, though ye once knew this, how that the Lord, having saved the people out of the land of Egypt, afterward destroyed them that believed not. 6 And the angels which kept not their first estate, but left their own habitation, he hath reserved in everlasting chains under darkness unto the judgment of the great day. 7 Even as Sodom and Gomorrha, and the cities about them in like manner, giving themselves over to fornication, and going after strange flesh, are set forth for an example, suffering the vengeance of eternal fire. 8 Likewise also these [filthy] dreamers defile the flesh, despise dominion, and speak evil of dignities. 9 Yet Michael the archangel [gluon], when contending with the devil [electricity, photon] he disputed about the body of Moses [the Gentiles], durst not bring against him a railing accusation, but said, The Lord rebuke thee. Jude 1:4-9 KJV

great Babylon came in remembrance before God, to give unto her the cup of the wine of the fierceness of his wrath.

20 And every island fled away,

> the photons and quarks and antiquarks that were added to the brains, and central nervous system to control or restrict movement in each body;

and the mountains

> deception embedded in the brain,

were not found.

21 And there fell upon men a great hail

> atoms, crystallized, stones.

out of heaven,

> the mind

every stone about the weight of a talent.[86]

ability to see, understand;

and men blasphemed God because of the plague of the hail;

atoms,

for the plague thereof

of crystallization, where every atom communicates with one God, Jesus Christ, Atom Lambda and sees everything as it is occurring,[87]

was exceeding great. Revelation 16:1-21 KJV

---

86 For the kingdom of heaven [the brain, the mind] is as a man travelling into a far country, who called his own servants, and delivered unto them his goods. 15 And unto one [Judah] he gave five talents [the ability to discern what should be done] using options from all five forces], to another [Israel] two [strong nuclear force, electromagnetism], and to another [the Z boson] one [whose talent comes from the electroweak force, who uses camouflage as a stratagem, to conceal, obscure intent]; to every man according to his several ability; and straightway took his journey. 16 Then he that had received the five talents went and traded with the same, and made them other five talents. 17 And likewise he that had received two, he also gained other two. 18 But he that had received one went and digged in the earth [to hid his motive, just as the Z boson did when he brought photon into proton so that he could rape gluon before the singularity, producing quarks and antiquarks], and hid his lord's money. 19 After a long time the lord of those servants cometh, and reckoneth with them. 20 And so he that had received five talents came and brought other five talents, saying, Lord, thou deliveredst unto me five talents: behold, I have gained beside them five talents more. 21 His lord said unto him, Well done, thou good and faithful servant: thou hast been faithful over a few things, I will make thee ruler over many things: enter thou into the joy of thy lord. 22 He also that had received two talents came and said, Lord, thou deliveredst unto me two talents: behold, I have gained two other talents beside them. 23 His lord said unto him, Well done, good and faithful servant; thou hast been faithful over a few things, I will make thee ruler over many things: enter thou into the joy of thy lord. 24 Then he which had received the one talent came and said, Lord, I knew thee that thou art an hard man [an atom], reaping [mass, quarks and antiquarks] where thou hast not sown, and gathering [the Gentiles] where thou hast not strawed: 25 And I was afraid, and went and hid thy talent [the electroweak force] in the earth [in neutron]: lo, there thou hast that is thine. 26 His lord answered and said unto him, [Z boson] Thou wicked and slothful servant, thou knewest that I reap where I sowed not, and gather where I have not strawed: 27 Thou oughtest therefore to have put my money to the exchangers [the weak force, W^3 boson], and then at my coming I should have received mine own with usury. 28 Take therefore the talent [the electroweak force] from him, and give it unto him which hath ten talents. 29 For unto every one that hath shall be given, and he shall have abundance: but from him that hath not shall be taken away even that which he hath. 30 And cast ye the unprofitable servant into outer darkness [the waste below the waist]: there shall be weeping and gnashing of teeth. Matthew 25:14-30 KJV

87 But the very hairs of your head are all numbered. Matthew 10:30 KJV

Revelation 17: Photon, the great whore of Babylon

1 And there came one of the seven angels which had the seven vials,

and talked with me,

synaptically;

saying unto me,

Come hither; I will shew unto thee the judgment of the great whore

that sitteth upon many waters:

reproducing quarks and antiquarks, bacteria, viruses, cancerous diseases, filthiness and people;

2 With whom the kings of the earth:

Judah, Israel and Jerusalem, the three wise men, and the Gentiles

have committed fornication,

in flagrante delicto, *manusturpare* ("to defile with the hand"), in association with *turbare "to disturb*;

and the inhabitants

bacteria, viruses, diseases

of the earth,

the human body,

have been made drunk with the wine

fecal matter and urine, the filthiness

of her fornication.

3 So he carried me away in the spirit

the light

into the wilderness

the mind is the wilderness:

And I saw a woman

photon

sit upon a scarlet-coloured beast,

a neutron particle, a Z boson, covered in iron, a bloody mess;

and upon photon is a cover

full of names of blasphemy,

having seven heads

the seven churches

and ten horns

ten lights, all lights, bosons, gluons and photons, of the Standard Model of particle physics.

4 And the woman was arrayed in purple

Tyrian purple, electricity for bruising

and scarlet colour,

blood with iron

and decked with gold

neurons

and precious stones

protons, gluons

and pearls

neutrons,

having a golden cup

the neuron-filled minds of people

in her hand

full of abominations[88]

unholy and profane, murderers of fathers and murderers of mothers, manslayers, whoremongers, for them that defile themselves with mankind, for menstealers, for liars,

88 But we know that the law is good, if a man use it lawfully; 9 Knowing this, that the law is not made for a righteous man, but for the lawless and disobedient, for the ungodly and for sinners, for unholy and profane, for murderers of fathers and murderers of mothers, for manslayers, 10 For whoremongers, for them that defile themselves with mankind, for menstealers, for liars, for perjured persons, and if there be any other thing that is contrary to sound doctrine; 11 According to the glorious gospel of the blessed God, which was committed to my trust. 12 And I thank Christ Jesus our Lord, who hath enabled me, for that he counted me faithful, putting me into the ministry; 13 Who was before a blasphemer, and a persecutor, and injurious: but I obtained mercy, because I did it ignorantly in unbelief. 1 Timothy 1:8-13 KJV

for perjured persons, blasphemers, persecutors and injurious to the weak, the innocent

and filthiness of her fornication:

consumption of fecal matter and urine; increasing electricity, charges, to die harder and more completely during orgasm. And death is imminent.[89]

5 And upon her forehead was a name written,

MYSTERY, BABYLON THE GREAT,

Who killed proton and raped gluon to produce her children, the Gentiles;

THE MOTHER OF HARLOTS

Those whose thighs are in constant motion through electromagnetic impulses

AND ABOMINATIONS OF THE EARTH

where reproductive organs are kept moist by the emulsification of fecal matter and urine.

6 And I saw photon, the woman drunken with the blood of the saints,

protons, Higgs bosons

and with the blood of the martyrs of Jesus:

neutrons, Z or W^3 bosons.

And when I saw her, I wondered

about her and looked upon her, his mother

with great admiration.

7 And the angel said unto me,

Wherefore didst thou marvel?

I will tell you the mystery of the woman,

and of the beast

mankind and the humankind, composed of energy, mass and light

that carries her

within their bodies[90]: for she, and her brood, are waste managers, waste removal; for

---

89 But of the fruit of the tree which is in the midst of the garden, God hath said, Ye shall not eat of it, neither shall ye touch it, lest ye die. Genesis 3:3 KJV

90 Which thou hast commanded by thy servants the prophets, saying, The land, unto which ye go to possess it, is an unclean land with the filthiness of the people of the lands, with their abominations, which have filled it from one end to another with their uncleanness. Ezra 9:11 KJV

toxic elimination

and their thoughts: for she is the teacher, introducer of all filthiness occurring by the vibrations of magnetism upon the thighs,[91]

which has the seven heads:

Ephesus, Smyrna, Pergamos, Thyatira, Sardis, Philadelphia, Laodicea;

and ten horns:

a complete set; all minds, thoughts in the wilderness that are heard as voices:

as Higgs boson, gluon, Z or W^3 boson and photons.

8 The beast that you saw:

the filthiness of mankind and humankind, the rebellion instigated by photon, that great whore,

was your ruler,

but is no more;[92]

and is not

alive; and may provide a moment of pleasure, but cannot give you life:

Eating photon's fruit, fecal matter and urine, may make you wise to the ways of photon, so that you are not tricked, but your insides will become rotten, foul smelling, as the fecal matter in the bowels;

Photon is generated every time we eat.

And, when we eat, photon

shall ascend out of the bottomless pit,

the bladder and bowels

and go into perdition:

a state of transformation, another form of energy

And they that dwell on the earth:

The Gentiles

Shall wonder,

91 Her filthiness is in her skirts; she remembereth not her last end; therefore, she came down wonderfully: she had no comforter. O LORD, behold my affliction: for [photon], the enemy hath magnified himself. Lamentations 1:9 KJV

92 Then will I sprinkle clean water upon you, and ye shall be clean: from all your filthiness, and from all your idols, will I cleanse you. Ezekiel 36:25 KJV

Like John wondered about his mother;

The Gentiles are

Those whose names were not written in the book of life,

in deoxyribonucleic acid, DNA, from period one, proton, the hydrogen atom,

from the foundation of the world,

when they behold the beast that was

their ruler,

and is not

alive,

and yet is

existing as part of the human story, a white light coming out of the eyes, the bladder and bowels.[93]

9 And here is to the mind which hath wisdom,

Jesus Christ, guiding it into truth.

The seven heads are seven mountains,

forebrains,

on which the woman

photon

sits,

and rules the thoughts, words, emotions and responses of all people.

10 And there are seven kings:

Every person who has a crown, the crown of the head,
Ephesus, Smyrna, Pergamos, Thyatira, Sardis, Philadelphia, Laodicea;

five are fallen

under photon's spell: lust, greed, emotional turmoil, insincerity, deceived and deceiving others;

and one

Philadelphia, a proton particle

Is:

---

93 Take heed therefore that the light which is in thee be not darkness. Luke 11:35 KJV

maintaining by separation from the masses;

and the other

Laodicea, a gluon particle

is not yet come

to rule the health of the body with antibody armies;

and when he comes, he must continue

to heal

for a short space.

11 And

the Gentiles, photon

the beast that was,

ruling over mankind, Judah, Israel and Jerusalem, like a petulant child;

and is not,

ruling by photon's whims, for the truth of his mysterious confidence is exposed;

even he is the eighth

king,

and is of the seven,

churches

and goeth into perdition

awaiting transformation, change from one form of energy to another, which will surely come by the three laws of thermodynamics.

12 And the ten horns

Voices, lights, bosons, gluons, photons

which thou sawest are ten kings,

all people, gametes, atoms, made possible by Jesus Christ,

which have received no kingdom:

no physical body, composed of energy, mass and light

as yet;

but they will receive power as kings,

from gluon soup to crystallized mass after

one hour

one minute

one second

with the beast:

electromagnetism in mass, the potential of visibility, within light, that is darkness, a photon particle.

13 These

gluon particles in protons and neutrons on the cusp of the reproductive system

have one mind:

to be conceived as people, living beings in this noisy visible world;

and shall give their power

their light

and strength

their energy

unto

the growing of mass,

the beast

from zygote to embryo to full term child.

Gluon, Eve, the mother of all.

14 These

children, born in this generation, the age of the Gentiles

shall

instinctively

make war with the Lamb,

Atom Lambda, Jesus Christ, the source and root cause of every life;

The well-spring, energy, mass and light, health and intuition;

and the Lamb

Jesus Christ

shall overcome them:

for he is Lord of lords,

ruler over all mass;

and King of kings:

ruler of all light, purveyor of understanding,

from his throne, the central nervous system;

and they that are with him

Higgs boson, gluon, the Z or W^3 boson

are called, and chosen,

from within proton, the hydrogen seas, the tree of life

and faithful;

to Jesus Christ, mediator of all energy, mass and light.

15 And he

the angel

saith unto me,

The

hydrogen

waters, which thou sawest, where

photon

the whore sitteth,

are peoples,

masses: protons, gluons with neutrons and the quark-antiquark pairs;

Judah, Israel and Jerusalem and the Gentiles;

and multitudes,

unified lights working together to perform good things or bad things upon Jesus Christ's command,

and nations,

those who follow Jesus Christ, having intuition;

those who follow photon, having instinct,

and tongues

truth, logic, emotions, deception.

16 And the ten horns

lights, that serve as voices

which thou sawest upon the beast,

composites of energy, mass and light, the mankind and the humankind

these shall hate

photon

the whore,

offering pleasure, electromagnetism, stimulation between the thighs

and shall make her desolate

unwanted, unneeded, in preference of gluon strong nuclear, ruler of reproduction

and naked,

exposing photon to the world, discussing her antics, comparing experiences;

and shall eat her flesh,

fecal matter and urine;

and burn her with fire

electricity.

17 For God hath put in their hearts

minds, thoughts

to fulfill his will

that all people should know photon and the effect of electromagnetism unleashed within the living body,

and to agree

that photon is merciless,

and give their kingdom

their bodies and minds, souls,

unto the beast,

photon ruling in the body,

until the words of God

Jesus Christ, Atom Lambda

shall be fulfilled.

18 And the woman which thou sawest is that great city,

Babylon, the reproductive organs, pleasure,

which reigns over the kings of the earth. Revelation 17:1-18 KJV

Revelation 18: Babylon is Fallen

1 And after these things, I saw

Michael, a gluon particle, mass and light,

another angel come down from heaven,

the sun,

having great power;

strong nuclear power

and the earth

all mass

was lightened with his glory

as gluon is mass, lit by the noble gas neon, which transcends all mass and light.

2 And he cried mightily with a strong voice, saying,

Babylon the great is fallen,

as ruler of people's thoughts;

is fallen,

as photon is thrown down, flicked away like dried mucous, from the sun;

the reproductive organs,

and is become the habitation of devils,[94]

photons

and the hold of every foul spirit,

quarks and antiquark;

and a cage

---

94 And when he was come to the other side into the country of the Gergesenes, there met him two possessed with devils, coming out of the tombs, exceeding fierce, so that no man might pass by that way. 29 And, behold, they cried out, saying, What have we to do with thee, Jesus, thou Son of God? Art thou come hither to torment us before the time? 30 And there was a good way off from them a herd of many swine feeding. 31 So the devils besought him, saying, If thou cast us out, suffer us to go away into the herd of swine. 32 And he said unto them, Go. And when they were come out, they went into the herd of swine: and, behold, the whole herd of swine ran violently down a steep place into the sea, and perished in the waters. 33 And they that kept them fled, and went their ways into the city, and told everything, and what was befallen to the possessed of the devils. Matthew 8:28-33 KJV

magnetism

of every unclean and hateful bird.

electricity;

The Gentiles, Ephesus carrying Sardis.

3 For all nations

Jewels, the Gentiles, those of the tree of life and the perpetrators of death;

have drunk of the wine,

urine

of the wrath

electricity,

of her, magnetism's, fornication,

with men and women, children and animals;

and the kings of the earth

Judah, Israel and Jerusalem and the Gentiles

have committed fornication with her,

with photon, electromagnetism;

and the merchants of the earth,

anyone who sells,

are waxed rich

in this age of the Gentiles

through the abundance of her

photon's

delicacies.

4 And I heard another voice from heaven,

W^3 boson

saying,

Come out of her,

out of photon, electromagnetism

my people, so that ye be not partakers of her sins,

deceptions; for she is deceiving you

and so that ye receive not of her plagues[95]

perpetuating plagues that lead to death and emotional turmoil for your offspring, and their generations by genetics.

5 For her sins have reached unto heaven,

the mind, affecting your mental health, snatching synapses on the verge of being revealed, your rational responses disappear in forgetfulness,

and God

Atom Lambda

hath remembered her iniquities.

6 Reward her,

photon electromagnetism

even as she rewarded you,

for following her into death, electricity, for a moment of pleasure,

she rewarded you with poverty, taking your joy and giving you worry

taking your children

taking your parents, friends and support;

and double unto her double

portion of misery

according to her works:

according to the misery you have suffered.

in the cup which she hath filled fill to her double.

Photon's legacy

7 How much she hath glorified herself,

Making claims on your life as if she is god;

and lived deliciously,

consuming your masses, eating food, nutrients, drinking blood causing anemia, eating masticated food offered to idols in the stomach, the colon; forces that maintain life,

95 Thou shalt not bow down thyself to them, nor serve them: for I the LORD thy God am a jealous God, visiting the iniquity of the fathers upon the children unto the third and fourth generation of them that hate me; Exodus 20:5 KJV

masses and light;

So, much torment and sorrow give her:

For she saith in her heart,

boasting:

I sit a queen,

The queen of heaven [96]

and am no widow

whose husband, electricity, is dead,

and shall see no sorrow.

8 Therefore, shall her plagues come in one day,

death,

electricity dies

and mourning,

the empathetic magnetism mourns the dead

and famine;

the wastes that provide photon's power are diminished

and she shall be utterly burned with fire:

the fire from the sun, Atom Lambda

for strong is the Lord God

Atom Lambda

who judgeth her.

9 And the kings of the earth,

Judah, Israel and Jerusalem and the Gentiles

who have committed fornication

---

96 The children [gluons] gather wood, and the fathers [Z or W^3 boson] kindle the fire, and the women knead their dough [in self-abuse], to make [yeast] cakes to the queen of heaven, and to pour out drink offerings [urine] unto other gods, that they may provoke me [Atom Lambda] to anger. 19 Do they provoke me to anger? saith the LORD: do they not provoke themselves to the confusion of their own faces? 20 Therefore thus saith the Lord GOD; Behold, mine anger and my fury shall be poured out upon this place, upon man, and upon beast, and upon the trees of the field, and upon the fruit of the ground; and it shall burn, and shall not be quenched. Jeremiah 7:18-20 KJV

while she guided their minds and hands into a moment of pleasure

and lived deliciously with her,

eating fecal matter and urine, contaminated blood wastes for the enjoyment of the women, who

shall bewail her, and lament for her, when they shall see the smoke of

electricity on the day of

her burning,

10 Standing afar off for the fear of her torment, saying,

Alas, alas, that great city Babylon, that mighty city! For in one hour is thy judgment is come.

11 And the merchants of the earth

the Z or W^3 boson, selling deception to the innocent, that serpent,[97]

shall weep and mourn over her; for no man buyeth their merchandise

lies

any more:

12 The merchandise of gold,

electricity

and silver, and precious stones, and of pearls,

knowledge, understanding and wisdom

and fine linen,

skin

97 Now the serpent [Z boson, photon and the W^3 boson unified in SU(3)] was more subtil than any beast of the field which the LORD God had made. And he said unto the woman, Yea, hath God said, Ye shall not eat of *every* tree of the garden? 2 And the woman said unto the serpent, We may eat of the fruit of the trees of the garden: 3 But of the fruit of the tree which is in the midst of the garden, God hath said, Ye shall not eat of it, neither shall ye touch it, lest ye die. 4 And the serpent said unto the woman, Ye shall not surely die: 5 For God doth know that in the day ye eat *thereof, then* your eyes shall be opened [to the pleasure that only comes with electricity], and ye shall be as gods, knowing good [waste elimination] and evil [a noticeable change to DNA]. 6 And when the woman saw that the tree [bladder and bowels] was good for food, and that it was pleasant to the eyes, and a tree to be desired to make one wise, she [kneeled down] took of the fruit thereof, and did eat [semen, containing fecal matter and urine], and gave also unto her husband with her; and he did eat. 7 And the eyes of them both were opened, and they knew that they were naked [unprotected from disease]; and they sewed fig leaves together [justifications], and made themselves aprons [to hide their thoughts]. Genesis 3:1-7 KJV

and purple,

veins

and silk,

bed linens

and scarlet,

blood

and all thyine wood, and all manner vessels of ivory, and all manner vessels of most precious wood, and of brass, and iron, and marble,

representing the involuntary state of blood engorged members for waste removal;

13 And cinnamon, and odours,

arising from the bowels

and ointments,

saliva

and frankincense,

calling photon aloud, without shame to fulfill their desire

and wine,

urine

and oil, and fine flour, and wheat,

yeast

and beasts, and sheep, and horses,

energy, mass and light

and chariots,

force-carrying particles

and slaves,

quarks and antiquarks;

and souls

particles of light

of men.

lost, without access to the one who saves.

14 And the fruits

fecal matter and urine

that thy soul

which is now photon

lusted after are departed from thee, and all things which were dainty and goodly are departed from thee, and thou shalt find them no more at all.

15 The merchants

Z or W^3 boson and Jerusalem, the human counterpart

of these things, which were made rich by her, shall stand afar off for the fear of her
torment, weeping and wailing, 16 And saying,

Alas, alas, that great city, that was clothed in fine linen, and purple, and scarlet, and decked with gold, and precious stones, and pearls!

17 For in one hour so great riches is come to nought. And every shipmaster, and all
the company in ships, and sailors, and as many as trade by sea, stood afar off, 18
And cried when they saw the smoke of her burning, saying,

What city is like unto this great city!

19 And they cast dust on their heads, and cried, weeping and wailing, saying,

Alas, alas, that great city, wherein were made rich all that had ships in the sea by reason of her costliness!

one moment of pleasure is equal to one life

for in one hour is she made desolate.

20 Rejoice over her,

demise

thou heaven, and ye holy apostles and prophets; for God hath avenged you on her.

21 And a mighty angel

gluon

took up a stone like a great millstone, and cast it into the sea, saying,

Thus with violence shall that great city Babylon be thrown down, and shall be found no more at all.

22 And the voice of harpers, and musicians, and of pipers, and trumpeters,

she sings, forcing people to sing with her; but she does not know the words to the song, but sings the same tune continuously, and it is difficult to get the tune out of your mind. It must be replaced with a new song. Now, her song

shall be heard no more at all in thee;

and no craftsman,

whether builder, knitter,

of whatsoever craft he be, shall be found any more in thee;

and the sound of a millstone

grinding, grinding

shall be heard no more at all in thee;

23 And the light of a candle

photon

shall shine no more at all in thee; and the voice of the bridegroom and of the bride

on their wedding night

shall be heard no more at all in thee: for thy merchants

those in SU(3),

were the great men of the earth; for by thy sorceries were all nations

the Jewels, the Gentiles

deceived.

24 And in her

Babylon, photon carrying electricity;

was found the blood of prophets, and of saints, and of all that were slain upon the earth. Revelation 18:1-24 KJV

Revelation 19: Babylon is Fallen

1 And after these things

I heard a great voice

one voice

of much people in heaven,

all people

saying, Alleluia; Salvation, and glory, and honour, and power, unto the Lord our God:

Atom Lambda, Jesus Christ, the Amen.

2 For true and righteous are his judgments: for he hath judged the great whore, which did corrupt the earth with her fornication, and hath avenged the blood of his servants at her hand.

3 And again they said, Alleluia. And her smoke rose up for ever and ever.

4 And the four and twenty elders and the four beasts fell down and worshipped God that sat on the throne, saying, Amen; Alleluia.

5 And a voice came out of the throne, saying,

Praise our God, all ye his servants, and ye that fear him, both small and great.

The Marriage of the Lamb, DNA to DNA

6 And I heard as it were the voice of a great multitude, and as the voice of many
waters, and as the voice of mighty thunderings, saying, Alleluia: for the Lord God
omnipotent reigneth. 7 Let us be glad and rejoice, and give honour to him: for the
marriage of the Lamb is come, and his wife

DNA

hath

been cleaned, stripped of her impurities and

made herself ready. 8 And to her was granted that she should be arrayed in fine
linen,

skin scrubbed well

clean and white: for the fine linen

fine skin

is the righteousness of saints.

9 And he saith unto me, Write,

Blessed are they which are called unto the marriage supper of the Lamb.

And he saith unto me,

These are the true sayings of God.

Atom Lambda

10 And I fell at his feet to worship him. And he

Jesus Christ

said unto me,

John

See thou do it not:

I am thy fellowservant,

of the same atom

and of thy brethren that have the testimony of Jesus: worship God: for the testimony of Jesus is the spirit of prophecy.

11 And I saw heaven opened,

in an atom

and behold a white horse;

gravity

and he that sat upon him was called Faithful

gluon

and True,

Atom

and in righteousness he doth judge and make war.

against the enemies of life.

12 His eyes were as a flame of fire,

the sun;

and on his head were many crowns

the crown of every head;

and he had a

scientific

name written, that no man knew, but he himself.

13 And he was clothed with a vesture

clothing, nerve and sinew

dipped in blood:

DNA

and his name is called

gravity

The Word of God.

he speaks to us and through us.

14 And the armies which were in heaven

every immunological army

followed him upon white horses,

with intelligence,

clothed in fine linen, white and clean.

15 And out of his mouth goeth a sharp sword,

electricity

that with it he should smite the nations:

and he shall rule them with a rod of iron:

the electroweak or weak force

and he treadeth the winepress

bladder and bowels

of the fierceness and wrath of Almighty God.

16 And he hath on his vesture and on his thigh a name written, KING OF KINGS,

every person whose crown of his head is visible.

AND LORD OF LORDS.

atomic particles with dominion.

17 And I saw an angel standing in the sun; and he cried with a loud voice, saying to all the fowls

photons, quarks and antiquarks

that fly in the midst of heaven,

the atom

Come and gather yourselves together unto the supper of

Atom Lambda

the great God; 18 That ye may eat the flesh of kings,

decomposing wastes, dead skin

and the flesh of captains, and the flesh of mighty men, and the flesh of horses, and of them that sit on them, and the flesh of all men, both free and bond, both small and great.

19 And I saw the beast,

electromagnetism

and the kings of the earth,

Judah, Israel and Jerusalem and the Gentiles

and their armies, gathered together to make war against him that sat on the horse, and against his army.

20 And the beast

magnetism

was taken, and with him the false prophet

electricity

that wrought miracles before him,

rescues

with which he deceived them that had received the mark of the beast,

choice, the ability to buy and sell their lives

and them that worshipped his image.

the phallic

These both were cast alive into a lake of fire

electricity

burning with brimstone

magnetism.

21 And the remnant

photon, electromagnetism

were slain with the sword

electricity

of him

Jesus Christ

that sat upon the horse,

gravity

which sword

electricity

proceeded out of his mouth:

and all the fowls

quarks and antiquarks

were filled with their flesh. Revelation 19:1-21 KJV

Revelation 20: Photon Chained in DNA

1 And I saw an angel come down from heaven, having the key of the bottomless pit

> the bladder and bowels

and a great chain

> of DNA reaching up to the beginning of time

in his hand.

2 And he laid hold on the dragon,

> photon,

that old serpent, which is the Devil,

> electricity sealed in a photon particle conjoined to

and Satan,

> magnetism sealed in a separate photon particle

and bound him a thousand years,

> a single day is a thousand years

3 And cast him into the bottomless pit,

> the bowels and bladder of every person

and shut him up,

> shut his mouth, for photon talks without ceasing unless you scream "shut up!" or sing a song that Jesus Christ gave you to sing;

and set a seal upon him,

> marking his separation from people so

that he should deceive the nations no more, till the thousand years[98] should be fulfilled:

and after that he must be loosed a little season.

> to avoid the explosion of gases magnetism in the bowels.

4 And I saw thrones,

> pelvic bones, where people sit

---

98 But, beloved, be not ignorant of this one thing, that one day is with the Lord as a thousand years, and a thousand years as one day. 2 Peter 3:8 KJV

and they sat upon them, and judgment was given unto them:

for how they lived their lives;

and I saw the souls of them

particles in the image they are designed

that were beheaded for the witness of Jesus,

he's us, Jesus Christ, all people;

and for the word of God,

ability to hear Jesus Christ in the mind[99] providing wise counsel

and which had not worshipped the beast,

photon

neither his image,

the phallic symbol

neither had received his mark

choice, the ability to choose

upon their foreheads,

designated as a follower of electromagnetism, death, rather than life

or in their hands;

which is to experience photon's pleasure

and they lived and reigned with Christ a thousand years.

one day

5 But the rest of the dead

photons

lived not again until the thousand years were finished.

This is the first resurrection.

of the lights that go out of the brain at death

6 Blessed and holy is he that hath part in the first resurrection: on such the second death hath no power, but they shall be priests

---

99 And they shall say to you, See here; or, see there: go not after them, nor follow them. Luke 17:23 KJV

those who listen

of God

Atom Lambda, the singularity

and of Christ,

Jesus Christ

and shall reign with him

Jesus Christ

a thousand years.

7 And when the thousand years are expired, Satan

photon

shall be loosed out of his prison, 8 And shall go out to deceive the nations which
are in the four quarters of the earth,

particles of life and death in an atom

Gog and Magog,

Gog is electricity, the imitation god

Magog is magnetism

to gather them

photon's brood,

together to battle: the number of whom is as the sand of the sea.

9 And they went up on the breadth of the earth,

up the back from the backside

and compassed

encompassed, surrounded

the camp

the organs of life

of the saints about, and the beloved city:

Jerusalem, where Jesus Christ rules in the hind brain

and fire

electricity

came down from God

Atom Lambda

out of heaven,

the mind

and devoured them.

10 And the devil

photon

that deceived them was cast into the lake of fire

electricity

and brimstone,

magnetism

where the beast

a photon

and the false prophet

electroweak or weak force

are, and shall be tormented day and night for ever and ever.

11 And I saw a great white throne,

the brain

and

Atom Lambda

him that sat on it, from whose face the earth

masses comprising the embodiment of every person

and the heaven

the mind

fled away; and there was found no place for them.

12 And I saw the dead, small and great, stand before God

Atom Lambda;

and the books

deoxyribonucleic acid, reaching up to the first man Adam; up Jacob [gluon's] ladder

were opened:

and all of the changes, additions and deletions may be seen; whether by prompted by people or by Jesus Christ;

and another book

the original blueprint from the singularity

was opened, which is the book of life:

and the dead were judged out of those things which were written in the books, according to their works.

recorded in DNA

13 And the sea

gravity

gave up the dead

photons

which were in it;

and death

electricity

and hell

magnetism

delivered up the dead which were in them:

quarks and antiquarks

and they were judged every man according to their works.

the promulgation and protection of life

14 And death

electricity

and hell

magnetism

were cast into the lake of fire.

This is the second death.

15 And whosoever was not found written in the book of life

written at the singularity, the beginning of all things visible

was cast into the lake of fire

the sun.

Revelation 20:1-15 KJV

Revelation 21: New Jerusalem: the Mind

1 And I saw a new heaven

new mind

and a new earth:

embodiment, the new anatomy of a person

for the first heaven and the first earth

the mind, diseased bodies in the living in this generation

were passed away; and there was no more sea.

mind tossing and turning, uncertainty.

2 And I John saw the holy city, new Jerusalem, coming down from God

the minds of all people coming

out of heaven, prepared as a bride adorned for

Jesus Christ

her husband.

3 And I heard a great voice out of heaven saying,

the brain

Behold, the tabernacle

the home

of God

Jesus Christ

is with men, and he

Jesus Christ

will dwell with them, and they

all people

shall be his people, and God

Atom Lambda

himself shall be with them, and be their God.

4 And God

Atom Lambda

shall wipe away all tears

misery

from their eyes;

and there shall be no more death, neither sorrow, nor crying, neither shall there be any more pain:

disease

for the former things

cares of life

are passed away.

5 And

Jesus Christ

he that sat upon the throne

at the midbrain

said, Behold, I make all things new. And he said unto me,

Write: for these words are true and faithful.

6 And he said unto me,

It is done. I am Alpha and Omega, the beginning and the end. I will give unto him
that is athirst of the fountain of the water of life freely. 7 He that overcometh

to the end

shall inherit all things; and I will be his God, and he shall be my son.

8 But the fearful, and unbelieving, and the abominable, and murderers, and whoremongers, and sorcerers, and idolaters, and all liars, shall have their part in the lake which burneth with fire and brimstone: which is the second death.

9 And there came unto me one of the seven angels which had the seven vials full of the seven last plagues, and talked with me, saying,

Come hither, I will shew thee the bride, the Lamb's wife.

10 And he carried me away in the spirit to a great and high mountain,

the brain

and shewed me that great city, the holy Jerusalem, descending out of heaven from
God, 11 Having the glory

light

of God

Atom Lambda, the sun:

and her light was like unto a stone

an atom

most precious, even like a jasper stone, clear as crystal;

12 And had a wall great and high,

to reduce and eliminate physical contact with the wastes that contaminate life, the wastes that put us in this state

and had twelve gates,

and at the gates, twelve angels,

gatekeepers

and names written thereon, which are the names of the twelve tribes of the children of Israel:

13 On the east three gates;

indicating three distinct versions of Israel, gluon particles;

on the north three gates;

indicating three distinct versions of the Gentiles, quarks and antiquarks particles;

on the south three gates;

indicating three distinct versions of Judah, proton particles, each with their purpose;

and on the west three gates.

indicating three distinct versions of Jerusalem, neutron particles, each with their purpose.

14 And the wall of the city had twelve foundations, and in them

was written

the names of the twelve apostles

every particle with a purpose

of the Lamb.

Atom Lambda

15 And he that talked with me had a golden reed to measure

the New Jerusalem

the city,

where Jesus Christ rules

and the gates

that keep photons, and the wastes with which they are embedded away from the living masses

thereof,

and the wall

to eliminate contact with wastes from the bladder and bowels.

thereof.

16 And the city

the mind

lieth foursquare,

one square for each particle of force

and the length

in the design of each force particle

is as large as the breadth:

a square

and he measured the city

the new Jerusalem, the minds of all people

with the reed, twelve thousand furlongs.

and none were lacking

The length and the breadth and the height of it are equal.

17 And he measured the wall

designed to keep out wastes

thereof,

and it was

an hundred and forty and four cubits,

> 144: one singularity, atom, four force-carrying particles and four forces: gravity, strong nuclear force, electroweak or weak force and electromagnetism

according to the measure of a man, that is, of the angel

> proton, who carries a gluon within for fornication.

18 And the building of the wall of it was of jasper:

> the wall of jasper is the outer nature of an atom

and the city

> within

was pure gold, like unto clear glass.

> like looking into the sun

19 And the foundations of the wall of the city were garnished with all manner of precious stones.

> force-carrying particles, placed in the original order in the book of life

The first foundation was jasper;

> proton

the second, sapphire;

> gluon

the third, a chalcedony;

> neutron

the fourth, an emerald;

> quarks and antiquarks

20 The fifth, sardonyx;

> proton

the sixth, sardius;

> gluon

the seventh, chrysolite;

> neutron

the eighth, beryl;

quarks and antiquarks

the ninth, a topaz;

proton

the tenth, a chrysoprasus;

gluon

the eleventh, a jacinth;

neutron

the twelfth, an amethyst.

quarks and antiquarks

21 And the twelve gates were twelve pearls;

of wisdom, a collection of experiences, of which we are reminded when required;

every several gate was of one pearl:

understanding, perception, to avoid conflicts, confusion

and the street of the city

Atom Lambda

was pure gold,

brilliant in its explanation, its ability to reveal to each of us what others cannot see

as it were transparent glass.

I get it!

22 And I saw no temple

within the atom; no place for people to worship, chanting long prayers

therein: for the Lord God Almighty and the Lamb are the temple of it.

and Atom Lambda already knows what you want to know and need

23 And the city had no need of the sun,

Atom Lambda provides understanding,

neither of the moon,

neutron is not needed

to shine in it: for

Atom Lambda,

the glory of God

the light of the sun

did lighten it, and

Jesus Christ

the Lamb is the light thereof.

24 And the nations of them which are saved

Judah, Israel and Jerusalem and Ephesus, Jewels and the Gentiles

shall walk in the light of it:

of Atom Lambda

and the kings of the earth

all people

do bring their glory

light and understanding

and honour into it.

25 And the gates of it shall not be shut at all by day: for there shall be no night

photons, neutrons carrying the electroweak force

there.

within the atom, the mind

26 And they

those who are saved

shall bring the glory and honour of the nations

each person representing the masses that are his kind

into it.

the atom, the mind

27 And there shall in no wise enter into it

the atom, the mind

anything that defileth, neither whatsoever worketh abomination, or maketh a lie:

photons;

but they which are written in the Lamb's book of life.

written from the foundation of the world.

Revelation 21:1-27 KJV

Revelation 22: Water of Life

1 And he

Jesus Christ

shewed me a pure river of water of life,

Jesus Christ, the river that goes out of Eden, the energy den, the brain, where the forces reside, and are commanded by Jesus Christ.

From thence it was parted, and became into four heads. Pison is Judah, gold; as proton is the basis for every atom, every life. Gihon is gluon, the helpmeet, which fortifies proton. Gihon, the strong nuclear force, encompasses the whole land of Ethiopia, the force of gravity.

And the third river is Hiddekel, the electroweak or weak force, where one is hidden and the face of the other is in plain view. This is Jerusalem the teacher; teaching good things and bad things. And the fourth river, Euphrates, electromagnetism, is part of SU(3); is hidden. This fourth river produced Ishmael, the seed of Abraham, the Gentiles, Sardis conjoined with Ephesus.

clear as crystal, proceeding out of the throne of God

Atom Lambda, residing in the sun;

and of the Lamb.

Jesus Christ, residing in the hindbrain.

2 In the midst of the street of it, and on either side of the river, was there the tree of life,

providing energy, mass and light

which bare twelve manner of fruits,

first fruit: Judah, Israel, Jerusalem, Ephesus

main crop: Judah, Israel, Jerusalem, Ephesus

final crop of the season: Judah, Israel, Jerusalem, Ephesus

and yielded her fruit every month:

today, who we are as crops, rotates every year;

and the leaves of the tree

celerity: Higgs boson, gluon, Z or W^3 boson

were for the healing of the nations.

the Gentiles and the Jewels.

3 And there shall be no more curse:[100]

Such as the one brought to bear by Adam and Eve, our ancestors;

but the throne of God

the sun;

and of the Lamb shall be in it;

the brain;

and his servants

proton, gluon and neutron, who are Judah, Israel, Jerusalem Z or W^3 boson and Ephesus, as people

shall serve him:

4 And they shall see his face;

for the face that everyone wears is his face: he's us crystallized, from zygote to death;

and his name

Jesus Christ

shall be in their foreheads.

minds.

5 And there shall be no night there;

no confusion concerning who or what the LORD is;

and they need no candle,

no photon to get people riled up;

neither light of the sun;

that is set above our heads;

for

Atom Lambda

the Lord God

and his son, Jesus Christ

---

100 And the LORD [Atom Lambda] smelled a sweet savour; and the LORD said in his heart, I will not again curse the ground any more for man's sake; for the imagination of [the blood flowing through a] man's heart is evil from his youth; neither will I again smite any more every thing living, as I have done. Genesis 8:21 KJV

giveth them light:

and they

Atom Lambda and Jesus Christ

shall reign for ever and ever.

until the end of spacetime.

6 And he said unto me,

These sayings are faithful and true:

and the Lord God

Atom Lambda

of the holy prophets sent his angel to shew unto his servants the things which must
shortly be done. 7 Behold, I come quickly: blessed is he that keepeth the sayings

instructions

of the prophecy of this book.

Do not deviate just because you know you are one of the chosen.

8 And I John saw these things, and heard them. And when I had heard and seen, I fell down to worship before the feet of

the W^3 boson, the teacher,

the angel which shewed me these things. 9 Then saith he unto me,

See thou do it not: for I am thy fellowservant, and of thy brethren the prophets, and of them which keep the sayings of this book:

worship God.

Atom Lambda

10 And he saith unto me,

Seal not the sayings of the prophecy of this book: for the time is at hand.

11 He that is unjust, let him be unjust still:

persecuting people;

and he which is filthy, let him be filthy still:

eating what should not be eaten;

and he that is righteous, let him be righteous still:

truthful;

and he that is holy, let him be holy still.

humble

12 And, behold, I come quickly; and my reward is with me, to give every man according as his work shall be.

13 I am Alpha

from gamete to zygote to embryo to foetus to birth

and Omega,

at death

the beginning and the end,

Jesus Christ, he's us is

the first and the last.

14 Blessed are they that do his

Atom Lambda's

commandments,

3 But of the fruit of the tree which is in the midst of the garden, God hath said, Ye shall not eat of it, neither shall ye touch it, lest ye die. Genesis 3:3 KJV

that they may have right to the tree of life, and may enter in through the gates into the city.

New Jerusalem

15 For without

outside

are dogs, and sorcerers, and whoremongers, and murderers, and idolaters, and whosoever loveth and maketh a lie.

photons, quarks and antiquarks, Sardis and Ephesus, and Jerusalem Z boson

16 I Jesus have sent mine angel

the W^3 boson

to testify unto you these things in the churches.

Ephesus the Gentiles, Smyrna as Judah the male, Pergamos as Israel the male, Thyatira as Jerusalem the male or female; Sardis the Gentiles, Philadelphia as Judah the female; and Laodicea as Israel the female;

Jesus Christ

am the root

proton

and the offspring of David,

gluon, Eve the mother of all, Israel, who delivered the gamete to the virgin, that was born without male sperm or fertilization

and the bright and morning star.

found in the hindbrain, of the cerebellar cortex; providing truth, knowledge, understanding and wisdom

17 And the Spirit

Jesus Christ

and the bride

the mind of every person

say,

Come.

home;

And let him that heareth say,

Come.

And let him that is athirst

for knowledge

come.

And whosoever will, let him take the water of life

that is energy, mass and light,

freely.

18 For I testify unto every man that heareth the words of the prophecy of this book,

the scriptures, from Genesis to Revelation:

If any man shall add unto these things, God

Atom Lambda

shall add unto him the plagues that are written in this book:

infirmities and death

19 And if any man shall take away from the words of the book of this prophecy,

changing the intent, the meaning,

God

Atom Lambda

shall take away his part out of the book of life,

DNA

and out of the holy city,

his mind, thoughts,

and from the things

promised

which are written in this book.

the scriptures

20 He

Jesus Christ

which testifieth

to

these things saith,

Surely I come quickly.

Amen.

This is for all men.

Even so,

come, Lord Jesus.

back to the brain, the body,

21 The grace of our Lord Jesus Christ be with you all.

Amen.

All men.

Revelation 22:1-21 KJV

APPENDIX

Code Words and Definitions

The Code Words were recognizable to me immediately. As soon as I saw a word I knew the value of the translation. It was English, common words, idioms.

Early on, when I first read Jesus Christ, I saw the name Cyrus. I could see it. "See you're us." Then I realized the code for the scriptures was written gently. There was no struggle to figure it out. It was handed to me.

Code Word	Definition
12,000	A coded message where the number 12 represents "all," and where the zeros in the formulation are only place holders without numerical significance, meaning nought, nothing.
A thousand years	A coded message for the number one with 1,000 years being as one day. The zeros in 1,000 are nought; have no meaning.
Adam	First man; son of Atom; created by Atom; DNA derived from Atom.
Aegis shield	The aegis, electricity on one side and magnetism on the other, is carried by Athena, the Higgs boson, and Zeus, Atom Lambda. The shield bears the head of a Gorgon, Medusa, electricity generating magnetism. "It produced a sound as from a myriad roaring dragon (Iliad, 4.17) and was borne by Athena in battle ... and among them went bright-eyed Athene, holding the precious aegis which is ageless and immortal: a hundred tassels of pure gold hang fluttering from it, electricity, tight-woven each of them, and each the worth of a hundred oxen." To be under the aegis is to be influenced by someone; it means doing something under the protection of a powerful, knowledgeable, or benevolent source.
Altar	The brain and heart, wherein blood and blessings flow.
Amen	A homophone, a-men, all men, all people.
Amnon	A homophone, "am none." As electricity, I am Sardis the Gentiles, no one. Because there is no oxygen in adenine, and no nucleus by the prokaryota standard, I am not sustained on my own. I die, convulsively, many times during a lifetime. And the evil ones live long lives with long memories, plotting revenge.
Apostle	He who listens to the voice, inclinations of Atom Lambda; a thinker, given to be a messenger; a person with the highest calling from Atom Lambda, Jesus Christ, sent forth to convey good news or to warn.
Atom Lambda	Lambda Λ; Life (energy, mass and light); is all things; mediates all things; is the visible image of Gravity, the invisible God; God in three persons, Higgs boson, gluon and the Z or W^3 boson; Atom Lambda is Λ Alpha, the first atom to makeup a gamete in the reproduction process and Ω Omega, the last atom to leave the body upon death.

Code Word	Definition
B boson	A light particle that carries phenomena of any of the fundamental interactions.
Babylon	A homophone, “Baby” refers to the reproductive system; “Lon” is indicative of the loins, the reproductive organs.
Beasts	A beast is a composite being of forces, mass and light, covered in muscle and skin; no longer mankind but living as human beings. Judah, Israel and Jerusalem and Cain are beasts. Judah, proton, Gravity, Higgs boson: consider the writing style of Matthew, Solomon, Jesus Christ. Like Gravity, Judah carries all forces, mass and light; his behaviour may appear as any of the four. Israel, gluon, strong nuclear, self-lit: consider the writing style of Mark, David, Jacob. Jerusalem, neutron, electroweak or weak force, Z or W^3 boson: consider the writing style of Luke and Paul, also called Saul and his mission as leader, teacher for the Gentiles, Esau. Cain, the Gentiles; quark-antiquark pairs, electromagnetism, photon: consider the writings of John, the life and actions of John the Baptist, Saul the king that was relentless in pursuing David, and Amnon, who raped and rejected Tamar.
Black hole	Waste organs where matter can exit but nothing should enter. A region of spacetime exhibiting such strong gravitational effects that nothing—not even particles and electromagnetic radiation such as light—can escape from inside it. General relativity predicts that a sufficiently compact mass, a hydrogen atom, can deform spacetime to form a black hole. The boundary of the region is called the event horizon, which has an enormous effect on the fate and circumstances of an object crossing it, However, no effects appear to be observed. A black hole is as an ideal black body, as it reflects no light. Its radiative temperature is on the order of a billionths of a kelvin for black holes of stellar mass, making it essentially impossible to observe. The anal cavity.

Code Word	Definition
Blood waste	Blood wastes are products of metabolism, deoxygenated blood carried away in hemoglobin. Arterial blood carries oxygen from inhaled air to all of the cells of the body and venous blood carries carbon dioxide. Hemoglobin is the main oxygen-carrying molecule in red blood cells and it carries both oxygen and carbon dioxide. Pulmonary arteries contain the most deoxygenated blood, while the pulmonary veins contain oxygenated blood.
Bottomless Pit	From the tongue to the bladder and bowels is the journey that food and drink takes. The bladder and bowels are Tartarus, the bottomless pit.
Bread	Fecal matter, feces.
Burn incense	Synaptic activity that lights up the forebrain and midbrain. Incense is the thoughts that occur to us and quickly fade away.
Chaldeans	The Chaldeans are antiquarks. Like chalk or clay, Ephesus, the magnetic side of electricity, crumbles easily, breaks down, unstable, having no atomic foundation.
Churches	Every person is a church. Every mind is a particle of light, a gem stones. People who think are churches. Thinking is prayer.
Conservation of energy, light, mass	In physics, the law of Conservation of Energy states that the total energy of an isolated system remains constant. In this way, it is said to be conserved over time. This means that energy, as it categorized in the form of the four universal forces, can neither be created nor destroyed; rather, it can be transformed only from its affected form of either energy or mass to another form, such as light.
Cooking Pot	The stomach.
Coulomb force	Coulomb's inverse-square law, is a law of physics that describes force interacting between static electrically charged particles. The force of interaction between the charges is attractive if the charges have opposite signs. It is repulsive if both have the same signed.

Code Word	Definition
Crop Rotation – 12-year	Pearls of wisdom for every several gates. In the harvest of the first fruits, jasper, sapphire, chalcedony and the emerald, Judah, Israel, Jerusalem and the Gentiles respectively, all receive pearls of wisdom from the start, special consideration, as they are chosen by Atom Lambda's own hand. In the main harvest, Judah, a sardonyx stone, Israel, sardius and Jerusalem, a chrysolite may each receive pearls of wisdom and overcome; however, the first of these are likely to be attractive, while those born during the main harvest are more likely to be repulsive. These are also chosen to be for a specified predestined purpose. In the final harvest of the 12-year season, the stragglers come through the uterus. Some are wise, and some are unwise. It will depend on the Z or W^3 boson, fate or fortune, the teacher and disciplinarian.
Cyclical Births by Proxy	In the days of the creation of Adam, proton and neutron, the pyrimidine bases of the three Jewels comprised a heterocyclic rotating ring of human births of Judah (proton, Higgs boson), Israel (gluon boson) and Jerusalem (neutron, the Z or W^3 boson). With the birth of Cain, the Gentiles were added to the cyclical rotation timeline. Since the birth of Cain, a cycle lasts, even now, for four periods (one year + for each force), three years each, in a hetero-generation cycle of 12.242375 months (or one year and six hours for 12 years).
Cyrus	Cyrus is a homophone: C-yr-us, See you're us. A message to me and you that we are all Cyrus. We are all atoms, energy, mass and light. The four Jewels are strong-Gravity, Cyrus. Cyrus is these four minds within strong-Gravity, separately recognized the Great (Judah as Gravity), the Elder (Jerusalem the weak-electroweak interaction), the Younger (Israel the strong nuclear interaction) and the Throne (the Gentiles, the usurpers, electromagnetism).
Daily sacrifice	The daily incremental elimination of waste, food not used for energy, from the bladder and bowels. 10 percent of food in the stomach is required for antiquark pairs, and photon carrying magnetism as they both live. Like electrons, electricity consumes waste from the large colon, the bladder and bowels.

Code Word	Definition
Devil	Electricity
DNA Backbone	The backbone of the DNA strand is made from alternating phosphate and sugar residues. The sugar in DNA is 2-deoxyribose, which is a pentose (five-carbon) sugar. The sugars are joined together by phosphate groups that form phosphodiester bonds between the third and fifth carbon atoms of adjacent sugar rings.
Dragon	Dragon that fought the woman: the three lights of the special unity group SU(3): Z boson, photon and the W^3 boson, otherwise known as the electroweak or weak force and electromagnetism combined as the electroweak force.
Drink offerings	Seminal fluid, mucous from the reproductive organs.
Eden	E-den or energy den; home of forces; also, body of man; also, Temple of God.
Edom	A homograph, e-dom, red, an electric domain or field.
Egypt	Egypt is e-gyp, energy gyp, where the forces within us sit stagnant all day waiting to be fed.
Elijah -1	Elijah is a prophet who lived before the time of Jesus Christ and provided instruction and understanding for the people.
Elijah -2	Elijah is e-lie-jah. Filled with the explosive energy of the electroweak force and the logic of the weak force, e-lie-jah is part of the Lie group. He carries the Z or W^3 boson, which is a group of subtly changing lights that make up Jerusalem's innate qualities. Pronounced /li:/ "Lee," a Lie group is a differentiable manifold, whose properties of the group operations are compatible with a smooth structure and symmetry found in continuous transformation groups. Lie groups are ubiquitous in mathematics and all areas of science. In this age, there are many stewards, Jerusalem, called Elijah. When Atom Lambda calls for Elijah he is calling for those who would teach his message with love like John and high respect like Paul and not be ashamed or afraid. If he calls you, you will know it. His love will be your reward for your Holy boldness.
Entropy	A measure of temperature, pressure, or composition of the energy that is not available for work during a thermodynamic process. A closed system, such as a person's body, evolves toward a state of maximum entropy, loss of heat equals death.

Code Word	Definition
Ephod	E-phod or energy pod; every atom is an energy pod; with the Higgs boson as the major energy pod, operating in a vacuum—the body and providing information and wisdom from the mind.
Eunuch	Quark-antiquark pairs, sterilized, unable to conceive, childless.
Euphrates River	Euphrates, You-freight-ease: Transport system for the digestive system from the tongue to the bladder and bowels. Was easy, like breathing; now digestion and elimination are difficult as photon and the waste management team are no longer doing their job. Remedies are required.
Evil spirit	Photon.
Feces, Faeces	Toxic Nature of Faeces: There is a difference between eating one's own faeces (autocoprophagia) and another's (allocoprophagia). Most human psychiatric reports refer to the former and not the latter. In the animal kingdom, coprophagia is also most commonly auto- or at least that of a relative most likely to share a similar microbiome. I suspect the dangers of transmitting enteric pathogens is greatest when they are not already part of one's own flora. Person-to-person transmission of Shigella, STEC, Hep A, Hep E and poliovirus are all well-known. For direct evidence, Shigella transmission via oral-anal route, probably provides the best-case series. Outbreaks, with this [oral anal eating] as a risk factor, and demonstrably identical biotypes and antibiograms [in gut flora], is about as good as you'll get.
Fertilization	The Z boson, the electroweak force, activates all gametes for fertilization. The gametes emerge as zygotes and grow into eggs. The egg is fertilized by its biological father, who renders his likeness, his features. In this way, Abram Z boson, is the father of all nations.
Fig leaves, clothing	Magnetism, lies, covering up true intentions with lies.
Fire of the altar	Synaptic activity, electricity.

Code Word	Definition
Flagellum	Type III secretion system: Found in Gram-negative bacteria, type III secretions are a result of divergent evolution. Phylogenetic analysis supports a model in which gram-negative bacteria can transfer the gene cassette horizontally to other species. The most researched of these are from a species of shigella (causes bacillary dysentery), salmonella (typhoid fever), Escherichia coli (gut flora, some strains cause food poisoning), vibrio (gastroenteritis and diarrhea), Burkholderia (glanders), Yersinia (plague), Chlamydia (sexually transmitted disease) and Pseudomonas (infects humans, animals and plants), including plant pathogens and the plant symbiont rhizobium.
Flesh	Used or oxygenated blood, matter.
Gentiles	Not a chosen people, grafted in to DNA beginning with Cain: born via rape of Eve in the fallopian tube as Tubal-Cain. She (electricity) conjoined with he (magnetism) is human, characterized by the conjoinment of electricity and magnetism, electromagnetism. The Gentiles are the conjoined human image of quark-antiquark pairs. They carry multiple personalities, up to 12. They are not auto-transitional like the three Jewels. The follow the laws of motion in that something has to physically stop them when they begin to tell lies to magnify themselves, murder, seek revenge, covet, steal, bear false witness, accuse, tempt and rebuke people to make themselves feel good. They appear as good people; however, by their origins, fecal matter and urine, their nature is foul.
Gluon boson, lord of the sugar rings	The 12 stones, each of which are called Christ sun of Man and is Christ Lambda, a hydrogen atom rising up as we grow, by way of helium, shall be engraved with the names of the children of Israel, the children of gluon are quark-antiquark pairs: up, down, strange, charm, top and bottom and their antiparticles. These 12 now appear within each of us three Jewels according to their names like the engravings of a sugar signet ring. Every Higgs boson has his own name cytosine, thymine, uracil, adenine and guanine, are set according to the twelve tribes born of gluon, also known as Jacob.

Code Word	Definition
God	Physics: Atom Lambda is God. From the singularity, he is the beginning of all things visible. By strong Gravity, Atom Lambda is the period one hydrogen atom, the Higgs boson—energy, mass and light. By Atom Lambda, everything is already planned. There are no coincidences. Make your plans and see if they come to fruition. If your plans are not in his plan it will never happen. You will be pleasantly surprised or deeply disappointed.
Gods	Forces: Gravity is the invisible God; with strong and weak forces and electromagnetism
Golden censer	A nerve cell, neuron.
Graven images	Reproductive organs, in which photons, quark-antiquark pairs and bacteria light to feed, in the grave, the bladder and bowels.
Great Red Dragon (GRD) Grand Unification	Grand Unification: Bosons, gluons and two photons linked together as one force mediated by Atom Lambda. The Great Red Dragon (GRD) is led by gluon strong nuclear the red horses; the electroweak or weak force, the black horses follow; Gravitation, the white horse, is third. Photon with electromagnetism is death. The grisled horse is electricity. The bay horse is magnetism. Together, they are the pale green horse, death. And electromagnetism is the least of these four and last; although, in this age, he appears to be first. The first, electromagnetism, shall be last and the last, Gravity, shall be first. (In the age of the Gentiles, Gravity is last in the unification. Without him, people have no reasoning.)
Grecia	Greek, like Dionysus, characteristically, whiling away the day with wine and pleasure.
Ground	Masticated food that has been put through strikes of electricity and flagella, a prokaryotic grinder in the stomach; where water has been removed for compression.
Heaven	The brain covered in synaptic light and the sun.
Heavy Chains	DNA chains, which are locked and unlocked at the will of Atom Lambda.

Code Word	Definition
Hebrew	A homophone, bosons and gluon lights that he, Atom Lambda, brewed like tea when the world was made. "He brews." Atom Lambda brewed the behaviours of his people as mass and light from the chemical elements that precede each noble gas during the six periods, the six days of creation. We are brewed, steeped, like a cup of tea. From crystallized elements, we Jewels are light, plasma, and particles of mass: protons, gluons, and neutrons. Five of the six Noble Gases, from periods one through three, five and six. The noble gas krypton, period four, is a deliberate manipulation, made by the Z boson, noble gas argon. Because photon, hence krypton is lethal to life, protons, only a Z boson could have made it.
Holy	Whole, uninjured, Latin sanctus, preserved as healthy, whole or intact, that cannot be transgressed or violated, regarded with happiness, good fortune.
Host	Particles and people are hosts to Atom in the hindbrain and the energy, mass and light that he commands within the kingdom, a person's body.
Human Condition	Being "human" is a property imposed on the Jewels with the grafting-in of Cain. By magnetism, we weep. We eat excessively. We drown our sorrows in strong drink. We are difficult to predict because of the errant forces, electroweak force, electricity and magnetism. We speak out of turn, loudly and would like to stop talking but cannot. We are being controlled by that spirit, photon. What is worse, we Jewels behave like the Gentiles behave: This is what it means to be human: Diseases passed from generation to generation, crude cursing, death, confusion, depression, psychological syndromes. All fallacies and misunderstandings are supplied by photon and his brood within you and his children the Gentiles. The death force is electromagnetism. Photon will kill you at Atom Lambda's command. Atom Lambda mediates everything that happens within your body, outside among your friends and enemies and within the purview of your workplace. The strong nuclear force is the death angel. Electroweak or weak force lured you into sleep.

Code Word	Definition
Hydrogen	By hydrogen, Atom Lambda, $C_6H_{10}O_4Po_4^{3-}$, we live. Strong-Gravity is the light at the center of the universe, the sun, the Higgs boson wherein resides the bosons of all people. We are the four living beings. We are the four universal forces as mass, as light, as viscera. We are hosts to the forces and their mass. The energy pods that we wear are hosts to the light. Their lights are our minds. By hydrogen we are Cyrus, a homograph, C-yr-us meaning See you're us!
Israel -1, a nation	Three Jewels led by Israel, gluon and the strong nuclear force.
Israel -2, a person	"Is real" is the visible image of the strong nuclear force; as a people is also Eve (of Adam and Eve); also, Jacob (Israel), and David.
Israel -3, a promised land	And Israel is "is real," the multitude of masses, of light, mass and matter particles, atoms that comprise every viscera, every living body. According a study from Washington University, there are approximately 100 trillion atoms in a human cell. Approximately the same number of cells are in the entire human body, meaning that 10 octillion atoms, or a 1 followed by 28 zeros, make up each person. *How many atoms are in a human cell? | Reference.com*
Jacob	Israel, literally "one that takes by the heel; a supplanter." (Genesis 25:26 KJV).
Jerusalem -1, Ham	Visible image of the weak force (teacher, serpent); from Abram to Abraham; pretender. Jerusalem the teacher rules all from the forebrain, the forehead, the two eyes. Ham is the last of Noah's three sons: Shem or Judah, Japheth or Israel and Ham as Jerusalem.
Jerusalem -2, an ideal	He (Atom Λ) rules all men via synapses fired from his throne in the cerebellum.
Jesus Christ	Jesus Christ is a homograph, Jes-us or He's us. Christ is short for crystalline, which connotes that we are crystallized from liquid hydrogen by way of condensate. Atom Lambda is indeed our ancestor, our Father, from period one periodic table of elements.

Code Word	Definition
Jesus Christ son of Man and the three major macromolecules	The five DNA nucleotides are joined one to one another in a chain by covalent bonds between the sugar of one nucleotide and the phosphate of the next, resulting in an alternating sugar-phosphate backbone. According to base pairing rules (A with T and C with G), hydrogen bonds bind the nitrogenous bases of the two separate polynucleotide strands to make double-stranded DNA. Each of the 12 archetypal foundation stones is associated with one of Atom Lambda's lights, proton carrying transitional lights: H, H+, H°, H^1, Qγ (magnetism), and gluon carrying auto-transitional lights: G°, G^1, γB (electricity), neutron carrying the auto-transitional lights of the Z or W^3 boson. The lights are separated according to their purpose of the lights. By the lights, we are chosen. By the lights we become as individuals, displaying specially honed character styles used to enhance the normal paragraph style that is Gravity, proton and Higgs boson.
Jewels: the chosen people	Jewels refer to method of creation; i.e. crystallization: Judah in the image of Gravity, (proton); Israel in the image of strong force (gluon); Jerusalem as weak force (neutron). The stones are atoms according to the periodic table of chemical elements as listed in the Revelation. Nine stones are set in a gold pyrimidine base, Atom Lambda's genome, the conjugated double bonds. These eight, constituting four male and four female hosts for strong-Gravity, strong nuclear, the electroweak-weak interaction, have gluon strong nuclear in their inclosings. The gametes are protons, gluons and neutrons. Note that neutron carries two lights within one gamete. He or she has a choice between cultivating a male or female mind, regardless of the sexual orientation of the energy pod.
Judah	Visible image of Gravity; Adam (of Adam and Eve); Sarai to Sarah; Solomon; Jesus Christ.
King	Ruler over a kingdom, empire of atoms; kings are those who wear crowns, the crown of the head.
Law of Conservation	Neither new types of mass nor new types of light may be created. What is here may not be destroyed. Everything is recycled. Life is restored on Atom Lambda's command to gluon, strong nuclear force.

Code Word	Definition
Law of Moses	Laws that were written to thwart by the laws of man, the natural inclinations of the Gentiles. These were never intended for the Jewels as Atom Lambda's laws are written in the blood. Rather, we are subject to universal laws, gravitational laws, laws of motion and the thermodynamic laws.
Linen	Integument, skin.
Lion's Den	E-den, energy den, where forces, lions, are mediated.
Man	Atom Lambda in the sun and carbon copies of Atom Lambda.
Mankind	Jesus Christ in the cerebellum, a carbon copy of Atom Lambda, which is literally the mind that rules the body of each living person.
Manna from heaven	Understanding concerning the previous day's observations.
Mass	The quantity of crystallized solids that a living body contains, as measured by its acceleration under a given force or by the force exerted on it by a gravitational field. Where, inertial mass measures an object's resistance to being accelerated by a force (F = ma); and active gravitational mass measures the gravitational force exerted by an object; and passive gravitational mass measures the force exerted on an object in a known gravitational field.
Mastery	The power of command, victory over the accusers, enemies of life.
Matter	A product of discarded wastes, (fecal matter and urine).
Melanin	A dark natural pigment found in the epidermis or skin adnexal structures. It is a complex polymer of oxidized tyrosine in response to actinic stimulation. It is bound to a carrier protein by melanocytes in the skin, mucous membrane, pia arachnoid, retina, inner ear and mesentery.
Most High	Atom Lambda is literally "most High" as he is a proton particle, which comprises the hydrogen atom that sits in the sun. He fuels life and growth by photosynthesis delivered in abundance from the sun.
Naked	Gravitational, truthful, honest, revealing their true feelings without shame.
Nebuchadnezzar	From Nebo: protect the boundary, otherwise known as strong nuclear, crystallized in a gluon particle.

Code Word	Definition
Nerve ending	The terminal structure of an axon that does not end at a synapse. An axon is a long nerve fiber that conducts away from the cell body of the neuron. A free nerve ending is not connected to any specific sensory receptor.
Neuron	A nerve cell, neuron that passes an electrical or chemical signal to another neuron or to the target efferent cell.
Oblation	A meal consisting of bread and wine.
Pauli Exclusion Principle	The Pauli exclusion principle is the quantum mechanical principle which states that two or more identical fermions (particles with half-integer spin, antisymmetric) cannot occupy the same quantum state within a quantum system simultaneously. Bosons with full integer spin are symmetric and are not subject to the Pauli exclusion principle, for any number of identical bosons can occupy the same quantum state as they are guided in their work by Jesus Christ, Atom Lambda's carbon copy on earth.
Peniel	The mind, the brain, a place for thinking, resolving issues: "for I have seen God, Jesus Christ, face to face, and my life is preserved." The face of Jesus Christ is each face. We see him face to face when we look at ourselves. (Genesis 32:30 KJV)
Penuel	The penis, rising of its own accord by photon: "And as he passed over Penuel, the flaccid penis, the sun, photon, rose upon him, and he, photon, halted upon Jacob's thigh." (Genesis 32:31 KJV)
Persia	The blackness of the gravitational field.
Photon in Jerusalem	Photon sits between the eyes, the third eye blind, directing the activities of your life. Photon is the center particle between the Z or W^3 boson in the forehead. In physics, this configuration is known as the special unitary group 3 or SU(3).
Photon, the ventriloquist	Photon is the original ventriloquist, a belly speaker, who speaks from the lower intestine. He causes sound to appear to be coming into the mind as thought from the cerebellar cortex, the brain. In this way, you may believe that Jesus Christ, the atom in the cerebellum, is speaking to you and through you. And he uses our voices to trick mankind, to trip us up.
Pleasant bread	Solid food masticated in the stomach for consumption by lights, bosons and gluons.

Code Word	Definition
Prayer	Thinking, Contemplation
Priest	One whose mind is a catalyst for reasoning, thinking, analyzing, understanding.
Priest	One that communes with Atom in the hindbrain by thinking, observing.
Promised Land	Refers to our bodies; comprised of atoms. As long as we follow the master -- Atom (light) we may live in the land -- our bodies.
Pulse	Food that has been chopped up, as in a blender in the stomach and passed as cooked food to the intestines, formatted for removal in the lower intestines so that it may be eliminated from the bowels through the anal cavity.
Purine	Pure urine, is a heterocyclic organic compound that is grafted into DNA through a pyrimidine ring fused to an imidazole ring. Purine gives its name to the wide class of molecules, which include substituted purines and corresponding tautomer, which is relevant to the behavior of amino acids and nucleic acids, two of the fundamental building blocks of life. In nature, purines occur most widely in nitrogen-containing heterocycles. Unlike pyrimidines, the dry mass, purine is water soluble. Purines are found in high concentration in meat and meat products, especially internal organs such as liver and kidneys, in beer (from the yeast) and gravy. In general, the plant-based diets, preferred by the Gentiles, are low in purines. There are two nucleobases derived from purine, adenine and guanine. An occurrence of guanine (Ephesus, magnetism) requires an occurrence of adenine (Sardis, electricity) and vice-versa. In DNA, these two bases form hydrogen bonds with thymine (Israel) and cytosine (Judah), respectively, from the days of Adam and Eve. In RNA, the complement of adenine is uracil (Abraham, father of all, instead of thymine (Eve, mother of all).

Code Word	Definition
Pyrimidine	A heterocyclic organic compound, where one of the three diazines (six-membered heterocyclics with two nitrogen atoms in the ring), has the nitrogen atoms at positions 1 and 3. Two other diazines are pyrazine (nitrogen atoms at the 1 and 4 positions). In nucleic acids, three types of nucleobases are pyrimidine derivatives: cytosine (Judah, DNA), thymine (Israel, DNA), and uracil (Jerusalem Z or W^3 boson, RNA).
Queen of Heaven	Photon carrying electricity conjoined with photon carrying magnetism conjoined as electromagnetism.
Redshift	A redshift occurs when the wavelength of light or other electromagnetic radiation of an object's mass is increased in wavelength or shifted to the red end of the spectrum. The result is equivalent to a lower, less precise frequency, a lower photon energy, in a wave or a thought.
Reprobate mind	And even as they did not like to retain God, Atom Lambda, in their knowledge, God, Atom Lambda, gave the Jewels over to a reprobate mind, to do those things which are not convenient under photon's rule. Now the Jewels are as human as the Gentiles, being filled with all unrighteousness, fornication, wickedness, covetousness, maliciousness; full of envy, murder, debate, deceit, malignity; whisperers. They became backbiters, haters of God, spiteful, proud, boasters, inventors of evil things, disobedient to parents. Because Jesus Christ, their light was withdrawn, the Jewels live without understanding, as covenant breakers, without natural affection, implacable, unmerciful, who knowing the judgment of God, ignore it. In this way, they can and do commit such things that are worthy of death, just as the Gentiles do. The Gentiles not only do all of the same things by their nature but have pleasure in the Jewels that do them and applaud their efforts to do evil. Romans 1:28-32 KJV
Righteousness	Operating in a wise manner. True to one's native self, according to the force and particles of mass and light in whose image a person is made. The righteous are genuine, excellent people whose foundation is the culmination of wise thought.
Saint	Endures for the sake of the body. Saints are mass, protons, gluons with neutrons and the quark-antiquark pairs, that follow instructions based on the choices to eat fecal matter and urine made by the hosts.

Code Word	Definition
Sanctuary	The mind, the outer courtyard extending from the forebrain up toward the radiating heat from the sun; and the body, living mass struggles in this age to survive.
Satan	Magnetism
Science	Science (12c.), from Latin, scientia, knowledge, expertness, intelligent, skilled, present participle of scire "to know"; having the ability "to separate one thing from another, to distinguish based on scientific conclusion while using logic in the philosophical sense with the objective of "considering the relation of the separate part to its impersonal, unbiased universal, physics and biological parts.
Sealed	To be sealed is to be made aware, to be taught on a certain subject or as a seed that is newly planted in the mind.
7, 10 and 12, 70	The numbers 7 and 10 are complete sets, a grouping. 12 is a cycle of events returning in its season.
Soul	Jesus Christ, the atom in the cerebellum that rules all energy, mass and light. He is you and me, every living person. Before birth, Jesus Christ becomes the husband to the bride: the boson or gluon light that gives you a name: Judah Higgs boson, Israel gluon and Jerusalem Z or W^3 boson. Bosons and gluons, lights are souls: Judah's soul is Higgs boson, Israel's soul is gluon strong nuclear, Jerusalem's soul is the Z or W^3 boson.
Soul-less	The Gentiles have no light, save photon, and are technically "soul less." However, some are chosen by Atom Lambda to marry themselves to Jesus Christ. Many in Ephesus follow him and some namely John also follow him with conviction.
Stones, Atoms:	Gametes for reproduction. The backbone of the DNA strand is made from alternating phosphate and sugar residues. The sugar in DNA is 2-deoxyribose, which is a pentose (five-carbon copies of the various composites of) sugar. The sugars are joined together by phosphate groups that form phosphodiester bonds between the third and fifth carbon atoms of adjacent sugar rings.
SU	Special unitary group.
SU(2)	Special unitary group, two forces.

Code Word	Definition
Temple of God	The temple at the forebrain beside the eye above the cheek bone. It is the region where we imagine that we are thinking, contemplating.
Thermodynamic Laws	The four laws of thermodynamics define fundamental physical quantities (temperature, energy, and entropy) that designate thermodynamic systems at thermal equilibrium between systems within a body of mass. The laws describe how mass behave under various circumstances, and how certain processes are forbidden.
Transgression of desolation	Violation of the second law of thermodynamics, which leads to invocation of the third law, zero energy, death.
Translation	In March 2008, my conscious mind was overtaken by a photon particle; it was a training session, where I was neurologically of unsound mind for some 24 hours. I had ample warning one week before and told Israel, the helpers, to bring me supplies and wait with me. They declined. In the morning, I found myself on the lawn in pajamas with bloody knees speaking in a monotone binary code, a language unknown to me. My son, Israel, who was given the role of rescuer in this bizarre lesson, rescued me by touching my hand and shouting at me. STOP TALKING! Only Thad could end this dreamlike state, as he is the only strong nuclear force among us. I was instructed by Atom Lambda to ask Thad if he understood what was happening to me. He said, “I do understand!” Then I was instructed to say, “Thank you.” Photon was spurred away from me and my mind resumed its normal state of operation.
Tree of Life	Atom’s fruit is light, life, joy and peace; planted in the cerebellum; Atom’s synaptic light is not detectable in death; omega Ω.
Tree of the knowledge of Good and Evil	Planted in the bowel; fruit is fecal matter and urine, which spreads seeds of lust, fear, hatred, terror, randomness, illness and death and obsession; magnetism confounds knowledge with emotion; electricity is death.
U(1)	A unitary group, one force. Gravity is one force; all forces reside within Gravity.
Ulai	The cerebellar cortex in every person; a homophone characterized by “U” or you, and “Lai” or lay, which means to rest in Gravity, in peace. Here Atom, which is each of us, rests and gives us rest, peace.

Code Word	Definition
Unification or Grand Unification	Mathematically, the unification is accomplished under the SU(2) × U(1) gauge group. The corresponding gauge bosons are the three W bosons of weak isospin from SU(2) ((W^3 = W1, W2 and W3). The B boson of weak hypercharge is derived of U(1), respectively, all of which are massless.
Where does the Higgs boson feed?	All light, including the Higgs boson, the Z and W^3 bosons and gluons, feed on the pleasant bread, the newly emulsified feast in the stomach. What we eat is provided as a masticated product in stage one of digestion. As purveyor of life, energy, mass and light within us, Jesus Christ, the Higgs boson, is fed first. Gluon is fed second. The Z boson and W^3 bosons are fed next. Photons consume the waste after it leaves the upper intestines. From the lower intestines, where waste is formed into logs, photons, quarks and antiquarks consume these to replenish their energy. These three also consume as they are processing undigested food into waste. This accounts for some 10 percent of shrinkage, called tithe. How can you stop the oxen pulling the plow from eating the grain?
Wine	Urine.
Worshipping the Image	The image, which was flaccid, is raised up by photon from the bowels and bladder. Eve was given to desire it and worship it on hands and knees. It is filled with photon's fruit, fecal matter and urine. Eve believed that swallowing Adam's seminal fluids heightened her pleasure; however, it was because photon magnetism was gnawing away on her genitals and a strike of electricity, which was the beginning of death for her, served to plant a quark-antiquark pair, a seed to bring Cain and his brother Able into the world, a real body, a robe, for a photon, death, to wear. In this way, photon came to walk among us. Electromagnetism begins the courtship ritual with a spark of electricity in the loins. Then the vibrations begin. Those who accept photon as lover, allowing magnetism to suck on the reproductive organs, it is worshipping other gods.

Predestination Schedule for Mankind by Universal Force

Based on the lunar regeneration of crops, this is a steadfast calendar, that marks the seasons for delivery by force, annually. When the system moves from annual to monthly, no one will know who is which crop in advance and it won't matter. For where there are two, Judah and Israel, or three Judah, Israel and Jerusalem together, Atom Lambda will be there among them. Your name says who you are. In my circle, all named John are the Gentiles. All whose names are David or Michael are Israel. Daniel and James are Jerusalem. Constance is Judah. Determine your own automatic code.

Mon	Day	Year	To	Mon	Day	Year	Force	Celerity	Crop
1	28	1645	—	2	15	1646	Gravity	Higgs boson	Judah
2	16	1646	—	2	4	1647	Electromagnetism	photon	Gentile
2	5	1647	—	1	24	1648	Electroweak or weak	Z or W^3 boson	Jerusalem
1	25	1648	—	2	10	1649	Strong nuclear force	gluon	Israel
2	11	1649	—	2	0	1650	Gravity	Higgs boson	Judah
2	1	1650	—	1	20	1651	Electromagnetism	photon	Gentile
1	21	1651	—	2	8	1652	Electroweak or weak	Z or W^3 boson	Jerusalem
2	9	1652	—	1	28	1653	Strong nuclear force	gluon	Israel
1	29	1653	—	2	16	1654	Gravity	Higgs boson	Judah
2	17	1654	—	2	5	1655	Electromagnetism	photon	Gentile
2	6	1655	—	1	25	1656	Electroweak or weak	Z or W^3 boson	Jerusalem
1	26	1656	—	2	12	1657	Strong nuclear force	gluon	Israel
2	13	1657	—	2	1	1658	Gravity	Higgs boson	Judah
2	2	1658	—	1	22	1659	Electromagnetism	photon	Gentile
1	23	1659	—	2	10	1660	Electroweak or weak	Z or W^3 boson	Jerusalem
2	11	1660	—	1	29	1661	Strong nuclear force	gluon	Israel
1	30	1661	—	2	17	1662	Gravity	Higgs boson	Judah
2	18	1662	—	2	7	1663	Electromagnetism	photon	Gentile
2	8	1663	—	1	27	1664	Electroweak or weak	Z or W^3 boson	Jerusalem
1	28	1664	—	2	14	1665	Strong nuclear force	gluon	Israel

Mon	Day	Year	To	Mon	Day	Year	Force	Celerity	Crop
2	15	1665	—	2	3	1666	Gravity	Higgs boson	Judah
2	4	1666	—	1	23	1667	Electromagnetism	photon	Gentile
1	24	1667	—	2	11	1668	Electroweak or weak	Z or W^3 boson	Jerusalem
2	12	1668	—	2	0	1669	Strong nuclear force	gluon	Israel
2	1	1669	—	1	20	1670	Gravity	Higgs boson	Judah
1	21	1670	—	2	8	1671	Electromagnetism	photon	Gentile
2	9	1671	—	1	29	1672	Electroweak or weak	Z or W^3 boson	Jerusalem
1	30	1672	—	2	16	1673	Strong nuclear force	gluon	Israel
2	17	1673	—	2	5	1674	Gravity	Higgs boson	Judah
2	6	1674	—	1	25	1675	Electromagnetism	photon	Gentile
1	26	1675	—	2	13	1676	Electroweak or weak	Z or W^3 boson	Jerusalem
2	14	1676	—	2	1	1677	Strong nuclear force	gluon	Israel
2	2	1677	—	1	22	1678	Gravity	Higgs boson	Judah
1	23	1678	—	2	10	1679	Electromagnetism	photon	Gentile
2	11	1679	—	1	30	1680	Electroweak or weak	Z or W^3 boson	Jerusalem
1	31	1680	—	2	17	1681	Strong nuclear force	gluon	Israel
2	18	1681	—	2	6	1682	Gravity	Higgs boson	Judah
2	7	1682	—	1	26	1683	Electromagnetism	photon	Gentile
1	27	1683	—	2	14	1684	Electroweak or weak	Z or W^3 boson	Jerusalem
2	15	1684	—	2	2	1685	Strong nuclear force	gluon	Israel
2	3	1685	—	1	23	1686	Gravity	Higgs boson	Judah
1	24	1686	—	2	11	1687	Electromagnetism	photon	Gentile
2	12	1687	—	2	1	1688	Electroweak or weak	Z or W^3 boson	Jerusalem
2	2	1688	—	1	20	1689	Strong nuclear force	gluon	Israel
1	21	1689	—	2	8	1690	Gravity	Higgs boson	Judah
2	9	1690	—	1	28	1691	Electromagnetism	photon	Gentile
1	29	1691	—	2	16	1692	Electroweak or weak	Z or W^3 boson	Jerusalem
2	17	1692	—	2	4	1693	Strong nuclear force	gluon	Israel
2	5	1693	—	1	24	1694	Gravity	Higgs boson	Judah
1	25	1694	—	2	12	1695	Electromagnetism	photon	Gentile
2	13	1695	—	2	2	1696	Electroweak or weak	Z or W^3 boson	Jerusalem
2	3	1696	—	1	22	1697	Strong nuclear force	gluon	Israel
1	23	1697	—	2	10	1698	Gravity	Higgs boson	Judah

Mon	Day	Year	To	Mon	Day	Year	Force	Celerity	Crop
2	11	1698	–	1	30	1699	Electromagnetism	photon	Gentile
1	31	1699	–	2	18	1700	Electroweak or weak	Z or W³ boson	Jerusalem
2	19	1700	–	2	7	1701	Strong nuclear force	gluon	Israel
2	8	1701	–	1	27	1702	Gravity	Higgs boson	Judah
1	28	1702	–	2	15	1703	Electromagnetism	photon	Gentile
2	16	1703	–	2	4	1704	Electroweak or weak	Z or W³ boson	Jerusalem
2	5	1704	–	1	24	1705	Strong nuclear force	gluon	Israel
1	25	1705	–	2	12	1706	Gravity	Higgs boson	Judah
2	13	1706	–	2	2	1707	Electromagnetism	photon	Gentile
2	3	1707	–	1	22	1708	Electroweak or weak	Z or W³ boson	Jerusalem
1	23	1708	–	2	9	1709	Strong nuclear force	gluon	Israel
2	10	1709	–	1	29	1710	Gravity	Higgs boson	Judah
1	30	1710	–	2	16	1711	Electromagnetism	photon	Gentile
2	17	1711	–	2	6	1712	Electroweak or weak	Z or W³ boson	Jerusalem
2	7	1712	–	1	25	1713	Strong nuclear force	gluon	Israel
1	26	1713	–	2	13	1714	Gravity	Higgs boson	Judah
2	14	1714	–	2	3	1715	Electromagnetism	photon	Gentile
2	4	1715	–	1	23	1716	Electroweak or weak	Z or W³ boson	Jerusalem
1	24	1716	–	2	10	1717	Strong nuclear force	gluon	Israel
2	11	1717	–	1	30	1718	Gravity	Higgs boson	Judah
1	31	1718	–	2	18	1719	Electromagnetism	photon	Gentile
2	19	1719	–	2	7	1720	Electroweak or weak	Z or W³ boson	Jerusalem
2	8	1720	–	1	27	1721	Strong nuclear force	gluon	Israel
1	28	1721	–	2	15	1722	Gravity	Higgs boson	Judah
2	16	1722	–	2	4	1723	Electromagnetism	photon	Gentile
2	5	1723	–	1	25	1724	Electroweak or weak	Z or W³ boson	Jerusalem
1	26	1724	–	2	12	1725	Strong nuclear force	gluon	Israel
2	13	1725	–	2	1	1726	Gravity	Higgs boson	Judah
2	2	1726	–	1	21	1727	Electromagnetism	photon	Gentile
1	22	1727	–	2	9	1728	Electroweak or weak	Z or W³ boson	Jerusalem
2	10	1728	–	1	28	1729	Strong nuclear force	gluon	Israel
1	29	1729	–	2	16	1730	Gravity	Higgs boson	Judah
2	17	1730	–	2	6	1731	Electromagnetism	photon	Gentile

Mon	Day	Year	To	Mon	Day	Year	Force	Celerity	Crop
2	7	1731	—	1	26	1732	Electroweak or weak	Z or W^3 boson	Jerusalem
1	27	1732	—	2	13	1733	Strong nuclear force	gluon	Israel
2	14	1733	—	2	3	1734	Gravity	Higgs boson	Judah
2	4	1734	—	1	23	1735	Electromagnetism	photon	Gentile
1	24	1735	—	2	11	1736	Electroweak or weak	Z or W^3 boson	Jerusalem
2	12	1736	—	1	30	1737	Strong nuclear force	gluon	Israel
1	31	1737	—	2	18	1738	Gravity	Higgs boson	Judah
2	19	1738	—	2	7	1739	Electromagnetism	photon	Gentile
2	8	1739	—	1	28	1740	Electroweak or weak	Z or W^3 boson	Jerusalem
1	29	1740	—	2	15	1741	Strong nuclear force	gluon	Israel
2	16	1741	—	2	4	1742	Gravity	Higgs boson	Judah
2	5	1742	—	1	25	1743	Electromagnetism	photon	Gentile
1	26	1743	—	2	12	1744	Electroweak or weak	Z or W^3 boson	Jerusalem
2	13	1744	—	2	0	1745	Strong nuclear force	gluon	Israel
2	1	1745	—	1	21	1746	Gravity	Higgs boson	Judah
1	22	1746	—	2	8	1747	Electromagnetism	photon	Gentile
2	9	1747	—	1	29	1748	Electroweak or weak	Z or W^3 boson	Jerusalem
1	30	1748	—	2	16	1749	Strong nuclear force	gluon	Israel
2	17	1749	—	2	6	1750	Gravity	Higgs boson	Judah
2	7	1750	—	1	26	1751	Electromagnetism	photon	Gentile
1	27	1751	—	2	14	1752	Electroweak or weak	Z or W^3 boson	Jerusalem
2	15	1752	—	2	2	1753	Strong nuclear force	gluon	Israel
2	3	1753	—	1	22	1754	Gravity	Higgs boson	Judah
1	23	1754	—	2	10	1755	Electromagnetism	photon	Gentile
2	11	1755	—	1	30	1756	Electroweak or weak	Z or W^3 boson	Jerusalem
1	31	1756	—	2	17	1757	Strong nuclear force	gluon	Israel
2	18	1757	—	2	7	1758	Gravity	Higgs boson	Judah
2	8	1758	—	1	28	1759	Electromagnetism	photon	Gentile
1	29	1759	—	2	16	1760	Electroweak or weak	Z or W^3 boson	Jerusalem
2	17	1760	—	2	4	1761	Strong nuclear force	gluon	Israel
2	5	1761	—	1	24	1762	Gravity	Higgs boson	Judah
1	25	1762	—	2	12	1763	Electromagnetism	photon	Gentile
2	13	1763	—	2	1	1764	Electroweak or weak	Z or W^3 boson	Jerusalem

Mon	Day	Year	To	Mon	Day	Year	Force	Celerity	Crop
2	2	1764	—	1	20	1765	Strong nuclear force	gluon	Israel
1	21	1765	—	2	8	1766	Gravity	Higgs boson	Judah
2	9	1766	—	1	29	1767	Electromagnetism	photon	Gentile
1	30	1767	—	2	17	1768	Electroweak or weak	Z or W^3 boson	Jerusalem
2	18	1768	—	2	6	1769	Strong nuclear force	gluon	Israel
2	7	1769	—	1	26	1770	Gravity	Higgs boson	Judah
1	27	1770	—	2	14	1771	Electromagnetism	photon	Gentile
2	15	1771	—	2	3	1772	Electroweak or weak	Z or W^3 boson	Jerusalem
2	4	1772	—	1	22	1773	Strong nuclear force	gluon	Israel
1	23	1773	—	2	10	1774	Gravity	Higgs boson	Judah
2	11	1774	—	1	30	1775	Electromagnetism	photon	Gentile
1	31	1775	—	2	18	1776	Electroweak or weak	Z or W^3 boson	Jerusalem
2	19	1776	—	2	7	1777	Strong nuclear force	gluon	Israel
2	8	1777	—	1	27	1778	Gravity	Higgs boson	Judah
1	28	1778	—	2	15	1779	Electromagnetism	photon	Gentile
2	16	1779	—	2	4	1780	Electroweak or weak	Z or W^3 boson	Jerusalem
2	5	1780	—	1	23	1781	Strong nuclear force	gluon	Israel
1	24	1781	—	2	11	1782	Gravity	Higgs boson	Judah
2	12	1782	—	2	1	1783	Electromagnetism	photon	Gentile
2	2	1783	—	1	21	1784	Electroweak or weak	Z or W^3 boson	Jerusalem
1	22	1784	—	2	8	1785	Strong nuclear force	gluon	Israel
2	9	1785	—	1	29	1786	Gravity	Higgs boson	Judah
1	30	1786	—	2	17	1787	Electromagnetism	photon	Gentile
2	18	1787	—	2	6	1788	Electroweak or weak	Z or W^3 boson	Jerusalem
2	7	1788	—	1	25	1789	Strong nuclear force	gluon	Israel
1	26	1789	—	2	13	1790	Gravity	Higgs boson	Judah
2	14	1790	—	2	2	1791	Electromagnetism	photon	Gentile
2	3	1791	—	1	23	1792	Electroweak or weak	Z or W^3 boson	Jerusalem
1	24	1792	—	2	10	1793	Strong nuclear force	gluon	Israel
2	11	1793	—	1	30	1794	Gravity	Higgs boson	Judah
1	31	1794	—	1	20	1795	Electromagnetism	photon	Gentile
1	21	1795	—	2	8	1796	Electroweak or weak	Z or W^3 boson	Jerusalem
2	9	1796	—	1	27	1797	Strong nuclear force	gluon	Israel

Mon	Day	Year	To	Mon	Day	Year	Force	Celerity	Crop
1	28	1797	—	2	15	1798	Gravity	Higgs boson	Judah
2	16	1798	—	2	4	1799	Electromagnetism	photon	Gentile
2	5	1799	—	1	24	1800	Electroweak or weak	Z or W^3 boson	Jerusalem
1	25	1800	—	2	12	1801	Strong nuclear force	gluon	Israel
2	13	1801	—	2	2	1802	Gravity	Higgs boson	Judah
2	3	1802	—	1	22	1803	Electromagnetism	photon	Gentile
1	23	1803	—	2	10	1804	Electroweak or weak	Z or W^3 boson	Jerusalem
2	11	1804	—	1	30	1805	Strong nuclear force	gluon	Israel
1	31	1805	—	2	17	1806	Gravity	Higgs boson	Judah
2	18	1806	—	2	6	1807	Electromagnetism	photon	Gentile
2	7	1807	—	1	27	1808	Electroweak or weak	Z or W^3 boson	Jerusalem
1	28	1808	—	2	13	1809	Strong nuclear force	gluon	Israel
2	14	1809	—	2	3	1810	Gravity	Higgs boson	Judah
2	4	1810	—	1	24	1811	Electromagnetism	photon	Gentile
1	25	1811	—	2	12	1812	Electroweak or weak	Z or W^3 boson	Jerusalem
2	13	1812	—	2	0	1813	Strong nuclear force	gluon	Israel
2	1	1813	—	1	20	1814	Gravity	Higgs boson	Judah
1	21	1814	—	2	8	1815	Electromagnetism	photon	Gentile
2	9	1815	—	1	28	1816	Electroweak or weak	Z or W^3 boson	Jerusalem
1	29	1816	—	2	15	1817	Strong nuclear force	gluon	Israel
2	16	1817	—	2	4	1818	Gravity	Higgs boson	Judah
2	5	1818	—	1	25	1819	Electromagnetism	photon	Gentile
1	26	1819	—	2	13	1820	Electroweak or weak	Z or W^3 boson	Jerusalem
2	14	1820	—	2	2	1821	Strong nuclear force	gluon	Israel
2	3	1821	—	1	22	1822	Gravity	Higgs boson	Judah
1	23	1822	—	2	10	1823	Electromagnetism	photon	Gentile
2	11	1823	—	1	30	1824	Electroweak or weak	Z or W^3 boson	Jerusalem
1	31	1824	—	2	17	1825	Strong nuclear force	gluon	Israel
2	18	1825	—	2	6	1826	Gravity	Higgs boson	Judah
2	7	1826	—	1	26	1827	Electromagnetism	photon	Gentile
1	27	1827	—	2	14	1828	Electroweak or weak	Z or W^3 boson	Jerusalem
2	15	1828	—	2	3	1829	Strong nuclear force	gluon	Israel
2	4	1829	—	1	24	1830	Gravity	Higgs boson	Judah

Mon	Day	Year	To	Mon	Day	Year	Force	Celerity	Crop
1	25	1830	—	2	12	1831	Electromagnetism	photon	Gentile
2	13	1831	—	2	1	1832	Electroweak or weak	Z or W³ boson	Jerusalem
2	2	1832	—	2	19	1833	Strong nuclear force	gluon	Israel
2	20	1833	—	2	8	1834	Gravity	Higgs boson	Judah
2	9	1834	—	1	28	1835	Electromagnetism	photon	Gentile
1	29	1835	—	2	16	1836	Electroweak or weak	Z or W³ boson	Jerusalem
2	17	1836	—	2	4	1837	Strong nuclear force	gluon	Israel
2	5	1837	—	1	25	1838	Gravity	Higgs boson	Judah
1	26	1838	—	2	13	1839	Electromagnetism	photon	Gentile
2	14	1839	—	2	2	1840	Electroweak or weak	Z or W³ boson	Jerusalem
2	3	1840	—	1	22	1841	Strong nuclear force	gluon	Israel
1	23	1841	—	2	9	1842	Gravity	Higgs boson	Judah
2	10	1842	—	1	29	1843	Electromagnetism	photon	Gentile
1	30	1843	—	2	17	1844	Electroweak or weak	Z or W³ boson	Jerusalem
2	18	1844	—	2	6	1845	Strong nuclear force	gluon	Israel
2	7	1845	—	1	26	1846	Gravity	Higgs boson	Judah
1	27	1846	—	2	14	1847	Electromagnetism	photon	Gentile
2	15	1847	—	2	4	1848	Electroweak or weak	Z or W³ boson	Jerusalem
2	5	1848	—	1	23	1849	Strong nuclear force	gluon	Israel
1	24	1849	—	2	11	1850	Gravity	Higgs boson	Judah
2	12	1850	—	2	0	1851	Electromagnetism	photon	Gentile
2	1	1851	—	2	19	1852	Electroweak or weak	Z or W³ boson	Jerusalem
2	20	1852	—	2	7	1853	Strong nuclear force	gluon	Israel
2	8	1853	—	1	28	1854	Gravity	Higgs boson	Judah
1	29	1854	—	2	16	1855	Electromagnetism	photon	Gentile
2	17	1855	—	2	5	1856	Electroweak or weak	Z or W³ boson	Jerusalem
2	6	1856	—	1	25	1857	Strong nuclear force	gluon	Israel
1	26	1857	—	2	13	1858	Gravity	Higgs boson	Judah
2	14	1858	—	2	2	1859	Electromagnetism	photon	Gentile
2	3	1859	—	1	22	1860	Electroweak or weak	Z or W³ boson	Jerusalem
1	23	1860	—	2	9	1861	Strong nuclear force	gluon	Israel
2	10	1861	—	1	29	1862	Gravity	Higgs boson	Judah
1	30	1862	—	2	17	1863	Electromagnetism	photon	Gentile

Mon	Day	Year	To	Mon	Day	Year	Force	Celerity	Crop
2	18	1863	—	2	7	1864	Electroweak or weak	Z or W^3 boson	Jerusalem
2	8	1864	—	1	26	1865	Strong nuclear force	gluon	Israel
1	27	1865	—	2	14	1866	Gravity	Higgs boson	Judah
2	15	1866	—	2	4	1867	Electromagnetism	photon	Gentile
2	5	1867	—	1	24	1868	Electroweak or weak	Z or W^3 boson	Jerusalem
1	25	1868	—	2	10	1869	Strong nuclear force	gluon	Israel
2	11	1869	—	1	30	1870	Gravity	Higgs boson	Judah
1	31	1870	—	2	18	1871	Electromagnetism	photon	Gentile
2	19	1871	—	2	8	1872	Electroweak or weak	Z or W^3 boson	Jerusalem
2	9	1872	—	1	28	1873	Strong nuclear force	gluon	Israel
1	29	1873	—	2	16	1874	Gravity	Higgs boson	Judah
2	17	1874	—	2	5	1875	Electromagnetism	photon	Gentile
2	6	1875	—	1	25	1876	Electroweak or weak	Z or W^3 boson	Jerusalem
1	26	1876	—	2	12	1877	Strong nuclear force	gluon	Israel
2	13	1877	—	2	1	1878	Gravity	Higgs boson	Judah
2	2	1878	—	1	21	1879	Electromagnetism	photon	Gentile
1	22	1879	—	2	9	1880	Electroweak or weak	Z or W^3 boson	Jerusalem
2	10	1880	—	1	29	1881	Strong nuclear force	gluon	Israel
1	30	1881	—	2	17	1882	Gravity	Higgs boson	Judah
2	18	1882	—	2	7	1883	Electromagnetism	photon	Gentile
2	8	1883	—	1	27	1884	Electroweak or weak	Z or W^3 boson	Jerusalem
1	28	1884	—	2	14	1885	Strong nuclear force	gluon	Israel
2	15	1885	—	2	3	1886	Gravity	Higgs boson	Judah
2	4	1886	—	1	23	1887	Electromagnetism	photon	Gentile
1	24	1887	—	2	11	1888	Electroweak or weak	Z or W^3 boson	Jerusalem
2	12	1888	—	1	30	1889	Strong nuclear force	gluon	Israel
1	31	1889	—	1	20	1890	Gravity	Higgs boson	Judah
1	21	1890	—	2	8	1891	Electromagnetism	photon	Gentile
2	9	1891	—	1	29	1892	Electroweak or weak	Z or W^3 boson	Jerusalem
1	30	1892	—	2	16	1893	Strong nuclear force	gluon	Israel
2	17	1893	—	2	5	1894	Gravity	Higgs boson	Judah
2	6	1894	—	1	25	1895	Electromagnetism	photon	Gentile
1	26	1895	—	2	12	1896	Electroweak or weak	Z or W^3 boson	Jerusalem

Mon	Day	Year	To	Mon	Day	Year	Force	Celerity	Crop
2	13	1896	—	2	1	1897	Strong nuclear force	gluon	Israel
2	2	1897	—	1	21	1898	Gravity	Higgs boson	Judah
1	22	1898	—	2	9	1899	Electromagnetism	photon	Gentile
2	10	1899	—	1	30	1900	Electroweak or weak	Z or W^3 boson	Jerusalem
1	31	1900	—	2	18	1901	Strong nuclear force	gluon	Israel
2	19	1901	—	2	7	1902	Gravity	Higgs boson	Judah
2	8	1902	—	1	28	1903	Electromagnetism	photon	Gentile
1	29	1903	—	2	15	1904	Electroweak or weak	Z or W^3 boson	Jerusalem
2	16	1904	—	2	3	1905	Strong nuclear force	gluon	Israel
2	4	1905	—	1	24	1906	Gravity	Higgs boson	Judah
1	25	1906	—	2	12	1907	Electromagnetism	photon	Gentile
2	13	1907	—	2	1	1908	Electroweak or weak	Z or W^3 boson	Jerusalem
2	2	1908	—	1	21	1909	Strong nuclear force	gluon	Israel
1	22	1909	—	2	9	1910	Gravity	Higgs boson	Judah
2	10	1910	—	1	29	1911	Electromagnetism	photon	Gentile
1	30	1911	—	2	17	1912	Electroweak or weak	Z or W^3 boson	Jerusalem
2	18	1912	—	2	5	1913	Strong nuclear force	gluon	Israel
2	6	1913	—	1	25	1914	Gravity	Higgs boson	Judah
1	26	1914	—	2	13	1915	Electromagnetism	photon	Gentile
2	14	1915	—	2	2	1916	Electroweak or weak	Z or W^3 boson	Jerusalem
2	3	1916	—	1	22	1917	Strong nuclear force	gluon	Israel
1	23	1917	—	2	10	1918	Gravity	Higgs boson	Judah
2	11	1918	—	2	0	1919	Electromagnetism	photon	Gentile
2	1	1919	—	2	19	1920	Electroweak or weak	Z or W^3 boson	Jerusalem
2	20	1920	—	2	7	1921	Strong nuclear force	gluon	Israel
2	8	1921	—	1	27	1922	Gravity	Higgs boson	Judah
1	28	1922	—	2	15	1923	Electromagnetism	photon	Gentile
2	16	1923	—	2	4	1924	Electroweak or weak	Z or W^3 boson	Jerusalem
2	5	1924	—	1	23	1925	Strong nuclear force	gluon	Israel
1	24	1925	—	2	12	1926	Gravity	Higgs boson	Judah
2	13	1926	—	2	1	1927	Electromagnetism	photon	Gentile
2	2	1927	—	1	22	1928	Electroweak or weak	Z or W^3 boson	Jerusalem
1	23	1928	—	2	9	1929	Strong nuclear force	gluon	Israel

Mon	Day	Year	To	Mon	Day	Year	Force	Celerity	Crop
2	10	1929	—	1	29	1930	Gravity	Higgs boson	Judah
1	30	1930	—	2	16	1931	Electromagnetism	photon	Gentile
2	17	1931	—	2	5	1932	Electroweak or weak	Z or W^3 boson	Jerusalem
2	6	1932	—	1	25	1933	Strong nuclear force	gluon	Israel
1	26	1933	—	2	13	1934	Gravity	Higgs boson	Judah
2	14	1934	—	2	3	1935	Electromagnetism	photon	Gentile
2	4	1935	—	1	23	1936	Electroweak or weak	Z or W^3 boson	Jerusalem
1	24	1936	—	2	10	1937	Strong nuclear force	gluon	Israel
2	11	1937	—	1	30	1938	Gravity	Higgs boson	Judah
1	31	1938	—	2	18	1939	Electromagnetism	photon	Gentile
2	19	1939	—	2	7	1940	Electroweak or weak	Z or W^3 boson	Jerusalem
2	8	1940	—	1	26	1941	Strong nuclear force	gluon	Israel
1	27	1941	—	2	14	1942	Gravity	Higgs boson	Judah
2	15	1942	—	2	4	1943	Electromagnetism	photon	Gentile
2	5	1943	—	1	24	1944	Electroweak or weak	Z or W^3 boson	Jerusalem
1	25	1944	—	2	12	1945	Strong nuclear force	gluon	Israel
2	13	1945	—	2	1	1946	Gravity	Higgs boson	Judah
2	2	1946	—	1	21	1947	Electromagnetism	photon	Gentile
1	22	1947	—	2	9	1948	Electroweak or weak	Z or W^3 boson	Jerusalem
2	10	1948	—	1	28	1949	Strong nuclear force	gluon	Israel
1	29	1949	—	2	16	1950	Gravity	Higgs boson	Judah
2	17	1950	—	2	5	1951	Electromagnetism	photon	Gentile
2	6	1951	—	1	26	1952	Electroweak or weak	Z or W^3 boson	Jerusalem
1	27	1952	—	2	13	1953	Strong nuclear force	gluon	Israel
2	14	1953	—	2	2	1954	Gravity	Higgs boson	Judah
2	3	1954	—	1	23	1955	Electromagnetism	photon	Gentile
1	24	1955	—	2	11	1956	Electroweak or weak	Z or W^3 boson	Jerusalem
2	12	1956	—	1	30	1957	Strong nuclear force	gluon	Israel
1	31	1957	—	2	17	1958	Gravity	Higgs boson	Judah
2	18	1958	—	2	7	1959	Electromagnetism	photon	Gentile
2	8	1959	—	1	27	1960	Electroweak or weak	Z or W^3 boson	Jerusalem
1	28	1960	—	2	14	1961	Strong nuclear force	gluon	Israel
2	15	1961	—	2	4	1962	Gravity	Higgs boson	Judah

Mon	Day	Year	To	Mon	Day	Year	Force	Celerity	Crop
2	5	1962	–	1	24	1963	Electromagnetism	photon	Gentile
1	25	1963	–	2	12	1964	Electroweak or weak	Z or W^3 boson	Jerusalem
2	13	1964	–	2	1	1965	Strong nuclear force	gluon	Israel
2	2	1965	–	1	20	1966	Gravity	Higgs boson	Judah
1	21	1966	–	2	8	1967	Electromagnetism	photon	Gentile
2	9	1967	–	1	29	1968	Electroweak or weak	Z or W^3 boson	Jerusalem
1	30	1968	–	2	16	1969	Strong nuclear force	gluon	Israel
2	17	1969	–	2	5	1970	Gravity	Higgs boson	Judah
2	6	1970	–	1	26	1971	Electromagnetism	photon	Gentile
1	27	1971	–	2	14	1972	Electroweak or weak	Z or W^3 boson	Jerusalem
2	15	1972	–	2	2	1973	Strong nuclear force	gluon	Israel
2	3	1973	–	1	22	1974	Gravity	Higgs boson	Judah
1	23	1974	–	2	10	1975	Electromagnetism	photon	Gentile
2	11	1975	–	1	30	1976	Electroweak or weak	Z or W^3 boson	Jerusalem
1	31	1976	–	2	17	1977	Strong nuclear force	gluon	Israel
2	18	1977	–	2	6	1978	Gravity	Higgs boson	Judah
2	7	1978	–	1	27	1979	Electromagnetism	photon	Gentile
1	28	1979	–	2	15	1980	Electroweak or weak	Z or W^3 boson	Jerusalem
2	16	1980	–	2	4	1981	Strong nuclear force	gluon	Israel
2	5	1981	–	1	24	1982	Gravity	Higgs boson	Judah
1	25	1982	–	2	12	1983	Electromagnetism	photon	Gentile
2	13	1983	–	2	1	1984	Electroweak or weak	Z or W^3 boson	Jerusalem
2	2	1984	–	2	19	1985	Strong nuclear force	gluon	Israel
2	20	1985	–	2	8	1986	Gravity	Higgs boson	Judah
2	9	1986	–	1	28	1987	Electromagnetism	photon	Gentile
1	29	1987	–	2	16	1988	Electroweak or weak	Z or W^3 boson	Jerusalem
2	17	1988	–	2	5	1989	Strong nuclear force	gluon	Israel
2	6	1989	–	1	26	1990	Gravity	Higgs boson	Judah
1	27	1990	–	2	14	1991	Electromagnetism	photon	Gentile
2	15	1991	–	2	3	1992	Electroweak or weak	Z or W^3 boson	Jerusalem
2	4	1992	–	1	22	1993	Strong nuclear force	gluon	Israel
1	23	1993	–	2	9	1994	Gravity	Higgs boson	Judah
2	10	1994	–	1	30	1995	Electromagnetism	photon	Gentile

Mon	Day	Year	To	Mon	Day	Year	Force	Celerity	Crop
1	31	1995	–	2	18	1996	Electroweak or weak	Z or W^3 boson	Jerusalem
2	19	1996	–	2	6	1997	Strong nuclear force	gluon	Israel
2	7	1997	–	1	27	1998	Gravity	Higgs boson	Judah
1	28	1998	–	2	15	1999	Electromagnetism	photon	Gentile
2	16	1999	–	2	4	2000	Electroweak or weak	Z or W^3 boson	Jerusalem
2	5	2000	–	1	23	2001	Strong nuclear force	gluon	Israel
1	24	2001	–	2	11	2002	Gravity	Higgs boson	Judah
2	12	2002	–	2	0	2003	Electromagnetism	photon	Gentile
2	1	2003	–	1	21	2004	Electroweak or weak	Z or W^3 boson	Jerusalem
1	22	2004	–	2	8	2005	Strong nuclear force	gluon	Israel
2	9	2005	–	1	28	2006	Gravity	Higgs boson	Judah
1	29	2006	–	2	17	2007	Electromagnetism	photon	Gentile
2	18	2007	–	2	6	2008	Electroweak or weak	Z or W^3 boson	Jerusalem
2	7	2008	–	1	25	2009	Strong nuclear force	gluon	Israel
1	26	2009	–	2	13	2010	Gravity	Higgs boson	Judah
2	14	2010	–	2	2	2011	Electromagnetism	photon	Gentile
2	3	2011	–	1	22	2012	Electroweak or weak	Z or W^3 boson	Jerusalem
1	23	2012	–	2	9	2013	Strong nuclear force	gluon	Israel
2	10	2013	–	1	30	2014	Gravity	Higgs boson	Judah
1	31	2014	–	2	18	2015	Electromagnetism	photon	Gentile
2	19	2015	–	2	7	2016	Electroweak or weak	Z or W^3 boson	Jerusalem
2	8	2016	–	1	27	2017	Strong nuclear force	gluon	Israel
1	28	2017	–	2	15	2018	Gravity	Higgs boson	Judah
2	16	2018	–	2	4	2019	Electromagnetism	photon	Gentile
2	5	2019	–	1	24	2020	Electroweak or weak	Z or W^3 boson	Jerusalem
1	25	2020	–	2	11	2021	Strong nuclear force	gluon	Israel
2	12	2021	–	2	0	2022	Gravity	Higgs boson	Judah
2	1	2022	–	1	21	2023	Electromagnetism	photon	Gentile
1	22	2023	–	2	9	2024	Electroweak or weak	Z or W^3 boson	Jerusalem
2	10	2024	–	1	28	2025	Strong nuclear force	gluon	Israel
1	29	2025	–	2	16	2026	Gravity	Higgs boson	Judah
2	17	2026	–	2	5	2027	Electromagnetism	photon	Gentile
2	6	2027	–	1	25	2028	Electroweak or weak	Z or W^3 boson	Jerusalem

Mon	Day	Year	To	Mon	Day	Year	Force	Celerity	Crop
1	26	2028	—	2	12	2029	Strong nuclear force	gluon	Israel
2	13	2029	—	2	2	2030	Gravity	Higgs boson	Judah
2	3	2030	—	1	22	2031	Electromagnetism	photon	Gentile
1	23	2031	—	2	10	2032	Electroweak or weak	Z or W^3 boson	Jerusalem
2	11	2032	—	1	30	2033	Strong nuclear force	gluon	Israel
1	31	2033	—	2	18	2034	Gravity	Higgs boson	Judah
2	19	2034	—	2	7	2035	Electromagnetism	photon	Gentile
2	8	2035	—	1	27	2036	Electroweak or weak	Z or W^3 boson	Jerusalem
1	28	2036	—	2	14	2037	Strong nuclear force	gluon	Israel
2	15	2037	—	2	3	2038	Gravity	Higgs boson	Judah
2	4	2038	—	1	23	2039	Electromagnetism	photon	Gentile
1	24	2039	—	2	11	2040	Electroweak or weak	Z or W^3 boson	Jerusalem
2	12	2040	—	2	0	2041	Strong nuclear force	gluon	Israel
2	1	2041	—	1	21	2042	Gravity	Higgs boson	Judah
1	22	2042	—	2	9	2043	Electromagnetism	photon	Gentile
2	10	2043	—	1	29	2044	Electroweak or weak	Z or W^3 boson	Jerusalem
1	30	2044	—	2	16	2045	Strong nuclear force	gluon	Israel
2	17	2045	—	2	5	2046	Gravity	Higgs boson	Judah
2	6	2046	—	1	25	2047	Electromagnetism	photon	Gentile
1	26	2047	—	2	13	2048	Electroweak or weak	Z or W^3 boson	Jerusalem
2	14	2048	—	2	1	2049	Strong nuclear force	gluon	Israel
2	2	2049	—	1	22	2050	Gravity	Higgs boson	Judah
1	23	2050	—	2	10	2051	Electromagnetism	photon	Gentile
2	11	2051	—	2	0	2052	Electroweak or weak	Z or W^3 boson	Jerusalem
2	1	2052	—	2	18	2053	Strong nuclear force	gluon	Israel
2	19	2053	—	2	7	2054	Gravity	Higgs boson	Judah
2	8	2054	—	1	27	2055	Electromagnetism	Photon	Gentile
1	28	2055	—	2	14	2056	Electroweak or weak	Z or W^3 boson	Jerusalem
2	15	2056	—	2	3	2057	Strong nuclear force	gluon	Israel